WORLD BANK TECHNICAL PAPER NUMBER 289

Restoring and Protecting the World's Lakes and Reservoirs

Ariel Dinar, Peter Seidl, Harvey Olem, Vanja Jorden, Alfred Duda, and Robert Johnson

The World Bank
Washington, D.C.

Manufactured in the United States of America
First printing July 1995

Technical Papers are published to communicate the results of the Bank's work to the development community with the least possible delay. The typescript of this paper therefore has not been prepared in accordance with the procedures appropriate to formal printed texts, and the World Bank accepts no responsibility for errors. Some sources cited in this paper may be informal documents that are not readily available.

The findings, interpretations, and conclusions expressed in this paper are entirely those of the author(s) and should not be attributed in any manner to the World Bank, to its affiliated organizations, or to members of its Board of Executive Directors or the countries they represent. The World Bank does not guarantee the accuracy of the data included in this publication and accepts no responsibility whatsoever for any consequence of their use. The boundaries, colors, denominations, and other information shown on any map in this volume do not imply on the part of the World Bank Group any judgment on the legal status of any territory or the endorsement or acceptance of such boundaries.

The complete backlist of publications from the World Bank is shown in the annual *Index of Publications,* which contains an alphabetical title list (with full ordering information) and indexes of subjects, authors, and countries and regions. The latest edition is available free of charge from the Distribution Unit, Office of the Publisher, The World Bank, 1818 H Street, N.W., Washington, D.C. 20433, U.S.A., or from Publications, The World Bank, 66, avenue d'Iéna, 75116 Paris, France.

ISSN: 0253-7494

Ariel Dinar is an economist in the Agricultural Policies Division of the World Bank's Agriculture and Natural Resources Department. At the time of writing, the other authors were in the Pollution and Environmental Economics Division of the Bank's Environment Department; Peter Seidl and Harvey Olem were consultants, Vanja Jorden was a research assistant, Alfred Duda was a senior environmental specialist, and Robert Johnson was an environmental specialist.

Library of Congress Cataloging-in-Publication Data

Restoring and protecting the world's lakes and reservoirs / Ariel
Dinar . . . [et al.].
p. cm. — (World Bank technical paper; no. 289)
ISBN 0-8213-3321-6
1. Water resources development. 2. Lakes—Management.
3. Reservoirs—Management. 4. Natural resources—Management.
I. Dinar, Ariel, 1947– . II. Series.
HD1691.R47 1995
333.91'63153—dc20 95–22293
CIP

Table of Contents

Page

Page

Boxes

Tables

Maps Section

Foreword

With the publication of its 1993 policy paper *Water Resources Management*, the World Bank made a commitment to assist developing countries in establishing institutional frameworks and management procedures that would enable countries to utilize their water resources in an economically and environmentally sustainable manner. The impetus for the policy came from the alarming deterioration and increasing scarcity of freshwater resources around the globe. In particular, substantial lake and reservoir degradation has resulted from ill-conceived and poorly executed development practices.

This paper calls attention to the particular problems that lakes and reservoirs have suffered as a result of increased siltation, nutrient loading, toxics, habitat degradation, over-exploitation, and changes to marginal ecosystems. These changes have destroyed fisheries needed to sustain large segments of the population, removed land from traditional farming and grazing practices, contaminated groundwater supplies, and reduced economic benefits. There are sensitive links between the quality of a lake or reservoir and development of its catchment. If these links are properly identified, it provides individuals in decision making positions opportunity for taking actions to protect vital lake and reservoir resources during the design and implementation of development projects.

The paper is intended to continue a process within the World Bank, collaborating with the United Nations System and the development community, giving priority for protecting these invaluable natural resources. As first steps, the Bank's 1993 policy paper and the 1994 technical paper *A Guide to the Formulation of Water Resources Strategy* both identify lake protection as an important resource management issue.

Our hope is that this paper, in combination with the aforementioned volumes will encourage professionals engaged in economic development to adopt sound environmental practices. It identifies vital components of the water-based ecosystem requiring special consideration to maintain biodiversity, including wetlands and marginal floodplains inextricably linked to lakes and reservoirs. If we engage in the process of developing our economies in a way that appropriately considers our natural resources, our lakes and reservoirs and watersheds, these resources will be protected for future generations to use and enjoy for all of their benefits.

Alexander F. McCalla, Director
Agriculture and Natural Resources Department

Andrew D. Steer, Director
Environment Department

We dedicate this paper to the memory of our colleagues Harvey Olem and Peter Seidl who were lost in an airplane crash in Bolivia in May 1994. Both Harvey and Peter had a special love for lake ecosystems. Harvey became an internationally recognized expert in reservoir water quality management as a result of many years of work on the Tennessee Valley Authority's vast reservoir system, and his work with Nordic scientists addressing acid precipitation effects. Peter's work over many years in the Canadian Civil Service brought him to the Canadian north to assist indigenous people coping with the construction of large hydroelectric reservoir systems. Concern with the native issues of native people then brought him to the North American Great Lakes and later to Lake Titicaca between Bolivia and Peru. In May 1994 Harvey and Peter were assisting the Bolivian government with a project to improve water resources management when their airplane was reported lost on the way to Lake Titicaca. The benefits provided to man and nature from lakes and reservoirs in both the North and the South are in danger. Some ecosystems are irreversibly damaged. Harvey and Peter recognized this and contributed to early drafts of this paper. In their memory, we hope that this call for action to protect the sustainable use of lake and reservoir ecosystems will receive the urgent attention it merits.

Acknowledgments
This paper benefited from comments provided to us by H. Kawaguchi and T. Kira, International Lake Environment Committee Foundation (ILEC); C. H. D. Magadza, University of Zimbabwe, R. Singh, R. Ramankutty, R. Crooks, and Colleagues, World Bank; D. LeMarquand, Ottawa, Canada; J. Jones University of Missouri-Columbia, D. Dudgeon, The University of Hong Kong; R E. Hecky, Government of Canada, R. Hirji, S. Lintner, C. Maxey, C. Rees, World Bank; and Walter Rast, UNEP, Nairobi, Kenya. Help with typing was provided by Maggie Wu. Cover Photographs of Lake Victoria, Kenya, were taken by Alfred M. Duda. The paper was copyedited by Sandra Giltner.

Abstract

Human activities are threatening the sustainable use of lakes and reservoirs around the globe. These sources of water are critical components in the ecological system. They provide habitat, sanctuary, and food for many species of fish and wildlife and are also a source of many beneficial uses for humankind. They provide drinking water; they are a source for process water to a myriad of industries, and their waters are used to generate power. Lakes and reservoirs provide food, navigation services, and recreational opportunities. Information and case studies reviewed in this paper indicate that the resource is at risk from over-enrichment, over exploitation, contamination by toxics, and water diversion for unsustainable agricultural uses. As a result, options for regional development are being curtailed, human populations are being uprooted, and future generations of people are being impacted by materials polluting lake waters. This paper calls for immediate action to reverse the degradation of these important natural resources. The paper argues that institutional processes should be developed to implement holistic, ecosystem-based, watershed-wide, stakeholder-driven strategies in order to sustain the resource. The way ahead was outlined by the World Bank in its Water Resources Management Policy adopted in 1993. The problems are so complex and the solutions require such a comprehensive approach that the World Bank, specialized UN agencies, and other international support organizations should place special emphasis on the problem. The paper calls on these organizations and governments to take necessary actions to make development activities more sustainable with regard to lakes and reservoirs.

Glossary

Aerobic	Describes life or processes that require the presence of molecular oxygen.
Algae	Small aquatic plants that occur as single cells, colonies, or filaments.
Algal bloom	A high concentration of a specific algal species in a body of water, usually caused by nutrient enrichment.
Alkalinity	A quantitative measure of water's capacity to neutralize acids. Alkalinity results from the presence of bicarbonates, carbonates, hydroxides, salts, and occasionally of borates, silicates and phosphates. Numerically, it is expressed as the concentration of calcium carbonate that has an equivalent capacity to neutralize strong acids.
Assimilation	The absorption and conversion of nutritive elements into protoplasm.
Anaerobic	Describes processes that occur in the absence of molecular oxygen.
Bioaccumulate	The process of biota to accumulate pollutants.
Biomass	The weight of biological matter. Standing crop is the amount of biomass (e.g., fish or algae) in a body of water at a given time. Often measured in terms of grams per square meter of surface.
Biota	All plant and animal species occurring in a specific area.
Blue-green algae	The phylum Cyanophyta, characterized by the presence of blue pigment in addition to green chlorophyll.
Chemical oxygen demand (COD)	Nonbiological uptake of molecular oxygen by organic and inorganic compounds in water.
Chlorophyll	A green pigment in algae and other green plants that is essential for the conversion of sunlight, carbon dioxide, and water to sugar. Sugar is then converted to starch, proteins, fats, and other organic molecules.
Chlorophyll-α	A type of chlorophyll present in all types of algae, sometimes in direct proportion to the biomass of algae.
Ecosystem	A system of interrelated organisms and their physical-chemical environment.
Embayments	Semi-enclosed shoreline areas of a lake or a reservoir.
Erosion	Breakdown and movement of land surface, which is often intensified by human disturbances.

Eutrophic — From Greek for "well-nourished," describes a lake of high photosynthetic activity and low transparency.

Eutrophication — The process of physical, chemical, and biological changes associated with nutrient, organic matter, and silt enrichment and sedimentation of a lake or reservoir. If the process is accelerated by man-made influences, it is termed cultural eutrophication.

Externality — The unintended real (nonmonetary) side-effects of one party's actions on another party that are often ignored in decisions made by the party causing the effect.

Flood plan — Land adjacent to lakes or rivers which is covered as water levels rise and overflow the normal water channels.

Hydrologic cycle — The circular flow or cycling or water from the atmosphere to the earth (precipitation) and back to the atmosphere (evaporation and plant transpiration). Runoff, surface water, groundwater, and water infiltrated in soils are all part of the hydrologic cycle.

Hypolimnion — The lower, cooler layer of a lake during summertime thermal stratification.

Macrophytes — Rooted or flooting aquatic plants.

Market Failure — A divergence between the market outcome (without intervention) and the economically efficient solution.

Opportunity cost — The value of goods or services foregone, including environmental goods and services, when a scarce resource is used for one purpose instead of for its next best alternative use.

Oxygen deficit — The difference between observed oxygen concentrations and the amount that would be present at 100 percent saturation at a specific temperature.

Oligotrophic — 'Poorly nourished," from the Greek. Describes a lake of low plant productivity and high transparency.

pH — A measure of the concentration of hydrogen ions of a substance, which ranges from very acid (pH = 1) to very alkaline (pH - 14). pH 7 is neutral and most lake waters range between 6 and 9. pH values less than 6 are considered acidic and most life forms can not survive at pH of 4.0 or lower.

Phytoplankton — Microscopic algae and microbes that float freely in open water of lakes and oceans.

Radionuclides — Radioactive subatomic particles.

Secchi depth	A measure of transparency of water obtained by lowering a black and white, or all white, disk (a Secchi disk, 20 cm in diameter) into water until it is no longer visible. Measured in units of meters or feet.
Sediment	Bottom material in a lake that has been deposited after the formation of a lake basin. It originates from remains of aquatic organisms, chemical precipitation of dissolved minerals, and erosion of surrounding lands (see ooze).
Stratification	Layering of water caused by differences in water density. Thermal stratification is typical of most deep lakes during summer. Chemical stratification can also occur.
Secchi depth	A measure of optical water clarity as determined by lowering a weighted Secchi disk into a water body to the point where it is not longer visible.
Suspended solids	Refers to the particulate matter in a sample, including the material that settles readily as well as the material that remains dispersed.
Terrace	An embankment or combination of an embankment and channel built across a slope to control erosion by diverting or storing surface runoff instead of permitting it to flow uninterrupted down the slope.
Trace elements	Those elements that are needed in low concentrations for the growth of an organism.
Urban runoff	Surface runoff from an urban drainage area.
Water quality standards	State-enforced standards describing the required physical and chemical properties of water according to its designated uses.
Watershed	A drainage area or basin in which all land and water areas drain or flow toward a central collector such as a stream, river, or lake at a lower elevation.
Zooplankton	Microscopic animals which float freely in lake water, graze on detritus particles, bacteria, and algae, and may be consumed by fish.

1
Introduction

Less than 3 percent of all water on the planet is fresh; of that, only 0.014 percent is readily available on the surface. Natural lakes of the world, on which hundreds of millions of people depend for their drinking water, contain more than 50 percent of this total. Lake Baikal and the Great Lakes in North America by themselves have more than 40 percent of the water contained in the earth's freshwater lakes.

Lakes have a more complex and fragile ecosystem than rivers. They do not have a "self cleaning" ability, and therefore they readily accumulate pollution. Because of their importance, their beauty, their religious and cultural significance, and their relative vulnerability to degradation, lakes and reservoirs require more concerted attention than is applied generally to river and stream basins. Although pollution of other waterways is also a pressing issue, the unique features and situation of lakes makes it important to deal with them first.

Sustainable use of lakes and reservoirs is being threatened around the globe, both in the North and in the South. Humanity depends on all sizes and kinds of lakes--natural or artificial, ephemeral or permanent--for drinking water, food supply, power production, navigation, ecological services, and economic development. Moreover, lakes are integrators of the many complex chemical, physical, and biological processes occurring throughout entire watersheds. They serve as indicators of adverse environmental change.

Over the last 50 years, the world population has almost tripled. Most of this increase has occurred in developing countries. The use of water during the same period has more than tripled. The exploitation of lands, bodies of water, and forests resulting from development has increased dramatically. This has led to a number of pronounced problems in lakes and reservoirs (hereafter mostly referred to as "lakes"). Changes in the natural landscape from resource extraction, introduction of new and more intensive types of agricultural practices, urbanization, altered water balances, and hydrological regimes have all influenced the state of lakes around the world.

Unfortunately, beneficial uses of lakes are being impaired, expensive investments in infrastructure are being lost, and ecological systems associated with lakes are collapsing. Substantial economic benefits from long-term, sustainable use of lake ecosystems will occur only if degraded lakes are restored and high-quality lakes are protected. If this is not done, permanent, irreversible damage can be expected.

The purpose of this paper is to alert policy leaders and decisionmakers to the importance of lakes and to demonstrate that the lakes and reservoirs of the world are at considerable risk. Continued loss of this resource jeopardizes sustainable economic growth. This paper describes lessons learned from lake degradation as well as from successful experiences in lake and reservoir rehabilitation, highlighting the importance of economic tools and sound financial policies. Processes need to be initiated to strengthen institutions, finances, and policies in order to reverse the continuing degradation and to establish improved approaches for lake and reservoir management.

The paper is neither a scientific treatise or a highly technical paper. Frequently, however, technical terms and concepts are presented to the general audience this paper addresses. Because of lack of consistent data needed to characterize the world-wide status and rate of change in the quality of lakes, the authors have drawn upon specific lake situations that, when considered collectively, characterize the degradation of lakes on a global scale.

The comprehensive approach to water resources management called for in the World Bank's 1993 water resources policy paper (World Bank 1993), is advocated here to protect lake and reservoir ecosystems. Conflicts can be resolved, international lake and reservoir catchments can be more effectively managed, pollution sources can be abated, and over-exploitation can be reduced. This paper calls attention to those issues countries need to address, and describes cases of successful practices for lake and reservoir management.

2
The Complex Interaction of Lake and Reservoir Ecosystems With Their Watersheds

There are many different types of lakes and reservoirs; and they serve many functions and purposes. Some lakes are ephemeral, such as floodplain lakes of large rivers. Others may be located in dry regions and are saline like the Aral Sea in Asia or the Great Salt Lake in America. Small saltwater lakes in coastal zones are essential for production of certain fisheries. Some lakes are recharged almost totally from groundwater inflow, while others, like lake Chad in Africa or small lakes in floodplains or delta areas in Asia, serve to recharge groundwater supplies. Depending on size, depth, climate, and flow regime, lakes differ in vulnerability to pollution.

There are significant differences between tropical and temperate lakes. In temperate lakes, nutrients are circulated during the autumn and winter, and the increase in temperature and illumination are the main triggers for seasonal primary production. Tropical lakes (once thought to show little seasonal primary production variances because of uniformly high temperatures and light throughout the year) also show seasonal variances. These are not as regular as in temperate lakes, and are mostly caused by excessive runoff during the rainy season, when large quantities of nutrients and organic material are carried to lakes, and during windy periods, when mixing is induced frequently to the bottom of shallower lakes. It is wind-induced mixing that causes the seasonal increase in production in Lakes Tanganyika, Malawi, and Victoria (Beadle 1981).

Thornton and Rast (1994) characterize the differences in temperate and tropical lakes by identifying differences in their watersheds. They note that the preponderance of lakes in the tropics are man-made. Watersheds of lakes in southern Africa were 2.5 times larger than those of similar northern lakes. They indicate that unit area loads of sediment, nutrients, and contaminants in southern Africa are similar to northern lakes, but because of the much larger watershed size, overall loading is much higher. This generally leads to bodies of water that are naturally turbid and rich in nutrients.

Lake and reservoir ecosystems contain a multitude of specialized subsystems such as wetlands, floodplains, and groundwater aquifers. Lakes continuously interact with their subsystems to obtain nourishment, inflow water, and a steady supply of organic matter and energy. This depends greatly on the management of their catchments (the geographical area from which the water flowing into a lake or reservoir originates, including the lake surface) and is linked to groundwater and aquifers.

Unfortunately, in the industrialized countries, lake ecosystems have been seldom considered in planning economic development, and a high cost is now being paid in the form of permanent ecosystem damage and expensive remedial action to restore at least some uses of contaminated and degraded systems. Because of the shortage of financial resources, and the need for immediate results, developing countries often do not consider lake--watershed interactions in their planning and projects. Consequently, the development may not be environmentally sustainable, and lake ecosystems on which people depend for their livelihood may be permanently damaged.

3

The Threat to Beneficial Uses of Lakes and Reservoirs

Lakes provide direct and indirect benefits to many segments of society. Lakes are also important ecological features that provide habitat for fish and shellfish, marginal nursery zones for many aquatic and amphibian species, resting places and feeding areas for numerous migratory birds, drinking water for countless species of mammals.

As a source of recreation, lakes support fishing, boating, and swimming. From a commercial point of view, lakes have considerable value in supplying food, supporting tourism, and providing a means of transportation. Lakes serve as flood control features and waste disposal sites. They are important components of hydropower generation, and a source of drinking, industrial, and irrigation water. In addition to these types of strictly economic benefits, people living around lakes demonstrate an affinity toward them. Such benefits are hard to quantify from an economic standpoint.

When valuing the benefits associated with a given lake, one should consider the relationships among the lake ecosystem and socio-economic components as demonstrated in the case of Lake Biwa, Japan (Box 1 and Annex 4). These relationships, which may vary over time and space, are dynamic, and may affect the result of an economic analysis. The relationships can vary according to the size of the lake, its location with respect to travel patterns, its placement in the landscape, its proximity to population centers, and other factors. The value lakes have to humans can and frequently does change in time as shown by Kada (1991) and Kira (1988a) for Lake Biwa. In prehistoric times, the greatest value of Lake Biwa was for production of food. In medieval and feudal times, this value remained important, but Lake Biwa was equally important for transportation. Today, the lake's most important uses are as a source of water, as a tool in flood control, and for both religious and leisure-time activities. All too frequently over-exploitation of a lake basin can increase pollution loading to a lake, which can trigger these changes in the uses and value of the lake.

It is clear that an economic analysis of beneficial uses associated with lakes should be comprehensive, dynamic, and include all potential uses and users since they affect each other. A special feature is the dynamic aspect of lake ecosystem degradation. The capacity of different types of lakes to sustain levels of pollution varies and is limited. In many cases, the damage to lake ecosystems, especially the biological communities, is irreversible. This means that benefits to local populations that depend on lakes are foregone. The poor may lose their source of nutrition, and others may lose their source of employment and may migrate. Furthermore, investments may be wasted. Many lake systems in the world are approaching limits or thresholds where their sustainability and beneficial uses will be severely diminished or lost. Corrective actions or costly remediation may be precluded if mismanagement and misuse of lakes and watersheds continue unchecked.

Box 1. Historical Changes in the Uses of Lake Biwa and in its Environment

Lake Biwa is 674 km^2 in area, and is located near the city of Kyoto (See Annex 4). In ancient times, Lake Biwa provided the local people with a variety of fish, shells, nuts, waterfowl and game that varied by season. With the introduction of agricultural practices such as irrigation, farmers started to use lake plants and lake mud as fertilizers. In addition, reeds and water plants were used for roofing, screens for housing, and gear for fishing. The role of the lake as a natural supplier of these raw materials declined, however, as new synthetic materials were introduced.

At the beginning of the century, water was pumped directly from the lake to near-by paddy fields, and a decade later, a textile industry that was established on the banks of the lake began to benefit from abundant amounts of good quality water. At the turn of the century, lake water was used for hydro-generation of electricity needed to operate street car systems, and for industrial purposes in neighboring cities. This role has declined since new dams were constructed in the nearby mountains in the 1960s. The lake itself serves as a natural flood control structure as it receives water from about 100 streams feeding it from its watershed.

Lake Biwa has served from ancient times as a route for transporting both people and goods. At the peak of transportation in the eighteenth century, as many as 3,000 boats traveled across the lake carrying goods produced in northern Japan to cities neighboring the lake. With the introduction of other means of transportation such as motor cars and electric trains, this role has vanished completely.

Lake Biwa represents a place of beauty and mystery for the people, and with the introduction of sight-seeing boats at the beginning of the 20th century, today about one million people annually enjoy Lake Biwa's beauty. Leisure use of Lake Biwa includes swimming, yachting, boating, sport fishing, and relaxing. Leisure activities diversified in the 1980s to include such activities as water skiing, and now more people arrive at the lake during holidays, increasing traffic jams, volumes of wastes, and noise, all of which are affecting the benefits associated with the lake. Finally, Lake Biwa has religious and symbolic meanings to several important sects of Buddhism that have built temples on its banks and islands.

Source: Kada (1991), (Kira 1988a).

To better dramatize the issue of beneficial uses of water, Endter (1987) compared cultural ideologies in values about water held by Northern Ute Indians and by Mormon ranchers in the United States. Although the purpose of Endter's study was to show attitudes towards water project development, it develops a basis for assigning non-market values to water when institutional and legal systems (courts, industry, and government) recognize only market and commodity valuations of water. Cultural differences towards values and beliefs regarding water may affect conservation and degradation of lakes. Box 2 presents the results of this study, including a listing of similarity measures of that describe the agreement and disagreement between the Ute and the Mormon cultures regarding nine water-related topics.

Several lessons can be drawn from Lake Biwa and from Endter's study of cultural differences. First, beneficial uses of lake water are interrelated and must be evaluated as a system. Second, the special nature of lake ecosystems necessitates careful decisions regarding timing and extent of correctional actions. Even with different views regarding relative importance of various beneficial uses by different segments of society, it is clear that in many cases the losses from precluded benefits may be larger than investments in restoration. This raises an additional question of whether conservation is more cost-effective than restoration. Finally, the various uses of lakes over time and across societies raises serious questions of equity that are associated both with pollution and with the methods to prevent it or to correct its consequences.

Box 2. Cultural Differences in Water Appreciation Between Utes and Mormons About Water-Related Topics

Topic	Extent of agreement (Similarity)
Spiritual or religious significance	10
Meaning of water	30
Buying and selling of water	34
Rules on use of water	60
Basis for rights to water	60
Importance of water	78
Loss of access to water	84
Consequences of markets for water	92
Practical uses of water	146

It is apparent that Mormon ranchers and Utes agree, but certainly not completely, on many practical uses of water. On other topics, however, both cultural and legal differences are dramatic. Values lower than 90 represent situations of disagreement, values between 90 and 100 represent non-significant relationships, and values exceeding 100 represent situations of agreement.

Source: Endter (1987).

Importance of small lakes, reservoirs, and ponds

Although it has never been documented, the number of small natural and man-made lakes, ponds, and reservoirs in the world (ranging from a few hectares to 200 hectares in surface area) is probably in the millions. For example, about 100,000 such lakes exist in the United States, excluding Alaska. These bodies of water offer the same range of important services to nearby urban and rural populations as large lakes and reservoirs do. Their accessibility, proximity to settlements, their extreme variety, and their shear numbers add dramatically to their importance. Unfortunately, small lakes are more susceptible to change by pollution than large lakes and many have lost their economic benefits even centuries ago as a result of unsustainable development around them.

Lakes Dobczyce and Goczalkowickie in Poland, for example, are the primary sources of drinking water for the 2 million residents of Cracow and Katowice, respectively. The two cities no longer can use the Vistula River as a drinking water source because of excessive industrial contamination . The lakes are jeopardized as a safe source of drinking water and industrial water supply, however, since they are becoming severely degraded by excessive nutrient and pathogenic inputs from agriculture (animal waste), small point sources, and untreated urban waste. As a result taste, odor, and algae problems are becoming serious. The situation has implications not only for the immediate health of the urban populations, but also for future development of the region; the latter will be precluded since funding to develop alternatives for domestic water resources is limited.

In Tanzania, Lake Mabayani, a man-made water supply reservoir, is filling rapidly with silt, jeopardizing the population of Tanga, a city of about 150,000 residents. The silt is coming from severe erosion problems in the Usambara Mountains as a result of overgrazing, cropping on very steep terrain, and over harvesting of wood for fuel supplies. If not stopped, this could lead to complete loss of the water source and a collapse of the socio-economic infrastructure of the region. Other freshwater sources or reservoir sites are not readily available.

Many smaller lakes have special ecological functions. One example is the oxbow lakes or lakes that lie in the floodplain areas of rivers in Asia, Africa, and South America. These lakes, subject to periodic flooding, serve as nurseries for many riverine fish species that provide the principal source of protein for local populations. Frequently, high silt loads, carried downstream from uncontrolled land use (such as intensive cropping on fragile soils, clear cutting of forest lands, or improper mining activities in the watershed) can fill oxbows, destroying vital habitats. Additionally, single-purpose operations of upstream dams for power production or irrigation can eliminate necessary floods down stream (see the example of Lake Manzala in Annex 5). The consequences of these interventions may include the migration of entire indigenous peoples to seek other sources of food and drinking water.

Protection and restorative measures

Over the past three decades, significant science and engineering progress has been made in understanding the links between lake basins and their watersheds. Methods have been perfected to characterize pollution loading sources by applying predictive models described by Cooke et al. (1993). Additionally, techniques have been developed to remedy degraded lakes, including the release of nutrients from bottom sediments to address internal recycling of nutrients and toxics, as well as to purge suspended algae and associated nutrients from the water column, and locking them in the bottom sediments. Dredging techniques have advanced to the point where little disturbance of the bottom sediments occurs and, re-suspension of any attached toxic materials is minimized (A complete review of contemporary lake management techniques is provided in Cooke et al. (1993) and in other literature published by the North American Lake Management Society).

As a result of these technical advances, watersheds of smaller lakes can be managed by significantly controlling pollution loading, and small lake basins can be manipulated to return many of their beneficial uses. Vancouver Lake, for example, a 1,200 hectare oxbow lake of the

Columbia River in the state of Washington in the United States, had for decades received excessive silt, chemical contamination from agriculture, toxics discharged from untreated stormwater, and insufficiently treated human waste. The ecological viability of the lake was vastly reduced and its economic value to the region had been almost completely lost. Studies of the lake and its watershed established the principal loading sources and defined appropriate measures to mitigate and correct these problems. Sources of pollution have now been controlled to satisfactory levels, and the removal of almost 700,000 cubic meters (m^3) of contaminated silt at a cost of more than $10 million has restored the lake as the centerpiece of economic development for the Port of Vancouver. Today, the lake is used for international sailing competition, sport fishing, swimming, and other forms of recreation.

Lake Boren and Lake Ekoln in Sweden demonstrate that highly eutrophic water bodies can be restored with the removal of excessive nutrient sources. Prior to 1970, these lakes received effluent discharges from municipal treatment plants providing only primary treatment of wastewater. Application of enhanced biological and chemical treatment to the wastewater reduced phosphorous loading to Lake Boren by 100 percent and to Lake Ekoln by 50 percent. In-lake chlorophyll concentrations were reduced by as much as 50 percent, and transparencies during the summer were improved by as much as 2 meters (m). Taste and odor problems from lake water have diminished as a result of shifting the algal composition away from blue-green to green algae.

Intergenerational considerations

Intergenerational effects may be classified in two types. The "immediate effects" result from a change in the natural resource base. The "cumulative effects" may not show for some time but can appear in later generations.

Immediate intergenerational effects may occur in several ways. Easy access to lake waters may be reduced, which results in greater costs and limits the amount of water availabile for future generations. Water diversions or changes in flow patterns upstream from a lake can have such immediate effects, which can also result in forced migration or in subtle changes in entire societies of poor people or indigenous peoples.

Excessive upstream water diversions have affected lake levels in the Aral Sea basin of Central Asia (Annex 9), the Tarim Lake basin of China, various delta lakes around the world (such as the Ganges or the Nile), and Lake Chad (Annex 7). In the latter case, diversions along with climatic shifts, have combined to lower the lake levels, reduce small lakes, inhibit access to groundwater, reduce fish populations, reduce access to lake water, and intensify the process of desertification. Similarly, in regulated river basins of Africa (such as the Senegal river basin) in which hydropower reservoirs have been created, whole biological communities and their associated ecosystems of ephemeral freshwater floodplain lakes have been destroyed by restricted reservoir outflows that inhibit annual flooding.

Even more pernicious are the cumulative intergenerational effects of persistent toxic chemicals that are released from industrial sites or waste dumps and that bioaccumulate to high concentrations in fish. People and aquatic animals that eat the fish may suffer subtle reproductive, behavioral, metabolic, immune system, and neurological problems. Some of these have been detected in scientific investigations of communities of people around the North American Great Lakes (Annex 10). Through bioaccumulation in food chains, these man-made chemicals act like artificial hormones to disrupt the endocrine systems of developing offspring. Persistent toxic substances mimic hormones and interfere with sexual development. Such toxics could be responsible for the more frequent occurrence of male and female reproductive system disorders and cancers found in humans in industrialized nations (Great Lakes Science Advisory Board 1991). Persistent toxic substances have been shown in epidemiological investigations in Michigan to be associated with impaired growth, impaired cognitive functioning, and behavioral problems in children born to mothers who consumed several Lake Michigan fish meals each month (Great Lakes Science Advisory Board 1991).

The Great Lakes experience shows that in industrialized areas, contamination of environment may also be damaging unborn children. This health problem goes to the very heart of pollution, equity, and intergenerational issues. Studies of these chemical effects on children have not been conducted in other countries as they have been in North America. If these effects are occurring in children around the Great Lakes, which are much cleaner than many of the lakes of the former Soviet Union, Eastern Europe, or rapidly industrializing countries of Asia, then intergenerational problems may be even more pervasive in other countries. Declining learning performance and behavioral problems in children, as well as cancers in adults, could be catastrophic worldwide in terms of human suffering, associated health costs, and indirect social costs such as foregone productivity. These effects are costly from a social and an economic perspective. Prevention of chemical pollution is much less costly in the long run than utilizing expensive remediation measures once a problem has been identified. Unfortunately, little progress is being made worldwide to diminish these risks.

Indigenous people and the poor

Worldwide, indigenous people and the poor people may suffer disproportionately from degradation of lake resources. In Lake Titicaca (Annex 6) the indigenous people have suffered the most from eutrophication and massive blooms of algae. The blooms shade and reduce the growth of the aquatic plants people need for food, fertilizer and building materials. On the shores of Lake Victoria (Annex 2), poor people used to catch abundant, inexpensive fish--almost 400 different species. Now, with the introduction of exotic fish species and with accelerated water pollution, one type of fish (an introduced species) dominates and its export value precludes its use as local food. On the shores of the second largest lake in the world, hunger has become a problem.

4
The Nature of Ecosystem Degradation

Lakes present special pollution problems because of their hydraulic character (Golubev 1989). Most notably, they serve as pollution sinks for upstream and lakeside activities including industrial discharges, municipal wastewater treatment, urban stormwater collection, and agricultural drainage. As a result, there are few lakes in the world that do not bear the imprint of human activity demonstrated by increased sedimentation, toxics contamination, and accelerated eutrophication. Most lakes serve many purposes, which greatly complicates strategies to protect or restore them. In addition, many lakes and their watersheds are shared by more than one political entity (e.g., provinces or countries), and this introduces political considerations that complicate the issues of water quantity and quality.

Lakes and reservoirs are part of larger ecosystems where, for instance, streams, springs, wetlands, and groundwater all interact. The problems in aquatic, terrestrial, and atmospheric ecosystems are linked. Stress in one part of the system frequently manifests itself as an adverse consequence in another. For example, the processes of nutrient recycling, sediment movement, fluctuations in rainfall, and anthropogenic emissions will over time alter ecosystems. Natural eutrophication or "aging" of a lake can take thousands of years, but the process can be radically changed and shortened by human intervention. This is particularly true with reservoirs. For example, Davy Crockett Lake behind Nolichucky Dam in Tennessee was completely filled with sediment from improper mining activities in its watershed within a few decades of building the dam. This lake was designed to provide beneficial uses for more than 100 years. The hydropower-generating capacity of the dam has been totally lost. The lake no longer serves as a water source for nearby communities, and fishing and other recreational activities have been vastly diminished. This situation is being replayed on an accelerated basis in Africa, South America, Asia, and in Central and Eastern Europe.

The protection of lakes and their restoration should be one of the primary aims of water management programs. Lake degradation problems can be corrected through expensive remedial actions, but preventing pollution through a variety of institutional and technological means is much more cost effective. The literature on this subject provides many examples of lake pollution problems, although very few are discussed in the context of economic, political, and institutional aspects. Well-known examples of lake water quantity and quality problems that also involve political aspects include the Great Lakes system, shared by two Canadian provinces and eight of the United States, and the the Aral Sea basin that is shared by five former Soviet republics. Several examples are provided in Box 3, and detailed information is presented in Annexes 1-12.

Box 3. Examples of Lake Pollution Problems

Lakes vary in the symptoms they display as a result of anthropogenic influences in their watersheds. The degree, extent and rate that lakes show signs of change are conditioned by many factors including basin size, water residence time, and the amount of pollution loading. The North American Great Lakes ecosystem, the largest freshwater system in the world, is characterized by a large human population, extensive industrial development, and intensified agriculture. The Great Lakes system is affected by eutrophication and toxic chemicals. Lake Balaton in Hungary also faces severe eutrophication. In Sweden, acidification affects 15,000 out of 85,000 lakes. Creation of artificial lakes in tropical countries, such as lake Cabora Bassa in Mozambique, contribute to the spread of water-born diseases such as malaria and schistosomiasis. Lake siltation is a problem in many developing countries, of which Lake Dal in Kashmir is one example. Over 13 million people depend upon Lake Biwa and its watershed for drinking water, agriculture, industry, and recreation. This largest lake of Japan and one of the oldest in the world has been a healthy source of fresh water, yet over the past three decades the lake has experienced a traumatic increase in pollution and a perilous slide in health. Hundreds of fish species have become extinct in Lake Victoria since the introduction of Nile perch and the eutrophication of its waters.

Source: Tolba (1984), Hecky (1993).

Reservoirs have certain features, not usually found with natural lakes, that make them even more difficult to protect, and once degraded, very difficult to improve. Reservoirs usually have very large drainage basins. Frequently they include extensive areas of agriculture that contribute large quantities of pollutants such as nonpoint nutrients, silt, and organic matter, making nutrient loading very high, and the probability of improvement in stream quality low. A single, high-volume source stream usually dominates the inflow; it may not only carry a heavy load of entrained pollutants and silt, but may also wash out reservoir restoration treatments such as chemical treatments of bottom sediments to reduce nutrient recycling, re-earation pumps floating on the surface, perforated hoses deployed on the reservoir bottom to oxygenate a reservoir.

Various ecological problems occur in lakes such as eutrophication, toxic contamination, accelerated sedimentation, acidification from atmospheric deposition, excessive water diversion, fish depletion, introduction of exotic species, and habitat alteration. A brief description of some of these ecological problems is provided below to demonstrate their variety, complexity, and geographical breadth.

Eutrophication

One of the most visible and common problems of lake ecosystems is accelerated enrichment or eutrophication. Eutrophication of a water body is a natural process. The original state of a lake and the rate at which eutrophication takes place depends on the location of the lake, is hydraulic character, and the attributes of its watershed including soils, slopes of terrain, and extent of vegetative cover. Lakes can be rather unproductive (oligotrophic), or highly enriched (hypereutrophic), or various degrees in between. Most lakes are eutrophic.

Concerns with eutrophication arise when anthropogenic activity changes the quality of a lake in a relatively short time. Increased eutrophy results from accelerated nutrient loading from expanded industrial activity, added wastewater discharges as in the case of Lake Washington (Box 4) and Lake Vanern (Annex 12), or expanded farming practices combined with sewage discharge as identified for Lake Titicaca (Annex 6). Increased eutrophication can also be caused by changing flow regimes through water diversion in the watershed, or modifying marginal environments by draining or covering wetlands that serve as natural nutrient traps.

The effects of accelerated eutrophication can include intensified algal blooms, which can create unsightly water conditions, unpleasant tastes and odors, and water discoloration. Some of these algae (the blue-greens) can produce toxins that are lethal to domestic animals and in temperate regions, and may be linked to certain summer illnesses in humans (USEPA 1988). Eutrophic conditions also contribute to the potential presence of toxic and harmful materials such as chlorinated organic hydrocarbons (e.g., trihalomethanes) in drinking water, which may persist even after water treatment. When algae die and decompose, dissolved oxygen concentrations in the lake bottom water can be depleted to the point of causing severe stress on fish. If the situation continues or intensifies, the result could be changes in species composition, growth reductions, and, under certain conditions, extensive fish death.

Macrophyte (rooted and flooting aquatic plants) abundance, while in part related to sediment type and composition, and to nutrient factors, is most often determined by light availability. It is a common misconception that nutrient enrichment of lakes causes nuisance macrophyte growth. In fact, high nutrient concentrations in the water column trigger algal blooms, decreasing light penetration, and limiting macrophyte growth (Cooke et al. 1993). Clear lakes with nutrient-rich littoral areas can display excessive macrophyte growth, which is very noticeable in Lake Chad and is apparent in Lake Victoria with its abundant growth of water hyacinths (see Annexes 7 and 2). Macrophyte beds are important features of lakes. They are essential as spawning areas for fish and shellfish, and provide refuge against predator species for some fish. They can provide important societal benefits as well, as in the case of Lake Titicaca, where certain species of macrophytes are used for nutrition and provide materials for construction as discussed in Annex 6.

Macrophyte beds become a problem when they cover large areas of a lake surface. Excessive growth can impede boat traffic, create dangerous areas to fisherman and swimmers, and can increase the cost of water treatment if plants break off and are transported to treatment facilities (where they can clog water intakes). The danger of introduced plant species (mostly unintentionally) is also a threat to lake sustainability. The invasion of North American water weeds (*Elodea spp.*) into other parts of the world is a well-known example. The distribution of *E. canadensis* is now almost world-wide, and *E. nuttallii* seems to follow a similar track, invading

Europe and Japan. Some lake ecosystems, for example those on oceanic islands, consist of a limited variety of flora and fauna and are less resistant to invader species.

Macrophyte decay, like that of algae, reduces the amount of oxygen produced, can over-enrich sediment, and can damage and disrupt fish habitat. Long-term control of plants most often involves deepening the lake in order to decrease available light. Control also often involves introducing animals that will eat exotic plants.

Box 4. eversing Rapid Eutrophication in Lake Washington, United States

Lake Washington is a long, narrow lake in an urbanized area. It forms the eastern border of the city of Seattle, Washington. It covers an area of approximately 88 km^2, with a volume of 3 x 10^9 m^3, and a maximum and mean depth of 65 and 33 meters respectively. The catchment area is rather small at about 1,300 km^2. The lake, offering a spectacular view of Mount Ranier, is used for swimming, fishing, sailing and other recreation.

Before 1964, Lake Washington was experiencing accelerated eutrophication. Total phosphorus concentrations had exceeded 64 micrograms (μg/l) as an annual mean, with chlorophyll (Ch α) concentrations reaching 36 μg/l. Wastewater entered Lake Washington from eleven small secondary treatment plants surrounding it. From 1963 to 1967 all eleven treatment plants, representing 88 percent of the external phosphorus loading, were diverted to a large primary treatment plant 3 km away, with effluent discharged at great depth in the Pudget Sound estuary.

As a result of the diversion, annual mean total phosphorus concentrations declined to an average concentration of about 21 μg/l and Chl α concentrations reached a seven-year (1969-1975) mean of 6 μg/l. Transparency increased threefold.

The Lake Washington success is probably the most celebrated lake recovery in the world. The success represented some of the first evidence of the singular importance of phosphorus to algal control. Also, the lake responded precisely as predicted by the models.

While the quick recovery of Lake Washington is not typical, lessons can be learned from it. The lake's large hypolimnion (lower layers), and its relatively short history of enrichment (first signs observed in the early 1950s) combined to prevent the hypolimnion from becoming oxygen-poor, making internal loading insignificant. Therefore, early treatment, as demonstrated by this lake, is an advantage. Even with this apparent success, however it would have been better if the waste loading had not been directed to the lake in the first place. In 1990 dollars, the construction cost for wastewater diversion was $366 million with an annual operating cost of $2.1 million.

Source. Cooke et al. (1993).

Toxic substances

In 1982, over 275 million tons of hazardous waste were produced by industry in 13 of the 23 Organization for Economic Cooperation and Development (OECD) countries that reported on waste disposal (Rogers et al. 1988). It is also estimated that 1.3 million metric tons of pesticides were used in the world in 1985. Although the amount of these substances that reach lakes and reservoirs is unclear, the shear size of the number indicates the potential for lake and reservoir pollution is very large. There is strong indication that a growth in hazardous waste production can be expected with increased industrialization and with production and use of synthetic organic compounds. Much of this expansion is to be expected in developing countries and more lakes and reservoirs will become contaminated. This potential growth in the production of hazardous waste is confirmed by the World Bank: in 1969 Thailand had only 500 factories producing hazardous waste, by 1992 that number grew to more than 26,000 factories, and it could triple in one more decade (World Bank 1992).

Toxic substances can enter waterbodies directly as land runoff from urban streets, mining areas, or as agricultural runoff including forestry drainage frequently including a range of toxics pesticides and herbicides, or by the discharge of insufficiently treated sewage and industrial effluents, and as deposition of airborne pollutants. Airborne transport is particularly troublesome since sources can be at great distances as indicated for the Great Lakes. In the case of Lake Michigan, the atmosphere is the main source of lead, zinc, copper, and cadmium. Toxics can also be created in drinking water when treatment plant chemicals interact with organic molecules in the raw water to form cancer-causing compounds such as trihalomethanes (THMs).

Harbors and coastal embayments, as identified in the Great Lakes Annexes 1, 10, and 11, are prone to building up these contaminants in bottom sediments. There, a range of heavy metals such as mercury, arsenic, lead, and cadmium as well as chlorinated substances such as DDT, DDE, PCBs, PAHs, and Dieldrin create toxic conditions. Toxics become available to organisms in the lower end of the food chain as they burrow and live in the sediments. Predator organisms higher in the food chain then bioaccumulate these materials to concentrations that may be as much as three hundred times that found in the ambient water (this has been reported for PCBs in the Great Lakes). The consumption by expectant mothers of fish heavily contaminated by toxics has previously been discussed as a major concern. When contrasted with the fact that in the 1970s, inland fisheries grew at a faster rate than those at sea (with the catches from lakes, streams, and ponds now making up nearly 15 percent of the aquatic harvests destined for human consumption), the concern for lake contamination by persistent toxics becomes even greater. In the last decade, freshwater aquaculture production grew at a yearly rate of about 7 percent compared with a 2 percent annual increase in tonnage of marine food fishes (Bardach 1988).

Sedimentation

Sedimentation of lakes is a normal occurance; in general, it takes place very slowly. Activities in a lake's watershed that disturb the soil can greatly accelerate this process, however. Most of these activities are human and include farming on fragile soils and on steep slopes, surface mining, forestry, construction activities, and urban drainage from streets and pavements.

Suspended sediment and sedimentation have many degrading effects on lakes and shoreline ecosystems. Marginal wetlands can be completely covered by silt, eliminating their value as nutrient sinks, wave absorbers, nursery areas, and habitat. As silt settles in a lake, spawning areas are covered, and lake volume is reduced. A complete change in fishery production can occur. As a result of lost storage capacity, both the volume and extent of flooding can increase. In the case of hydro-electric facilities, loss of generating capacity is possible. The increase in shoal areas as a result of sedimentation can encourage increased macrophyte growth, and can interfere with recreational activities such as boating and fishing. Soil loss and suspended silt also contributes to eutrophication and lake contamination since the silt generally includes attached nutrients, herbicides, pesticides, and other chemicals. This increases water treatment costs and maintenance problems in water treatment plants.

Sediment removal might be the only practical way to improve a lake or reservoir that has been badly silted. Dredging is one of the most frequently prescribed techniques to correct this problem. It has been successfully used in Lake Trummen, a small lake in southern Sweden, and in Lake Vancouver in the United States. Dredging, however, is costly (for example, Hitzhusen et al. (1995) estimated the dredging cost in Lake Madison in Ohio to be $1.65 per ton of sediments dredged) and works best if the sources of sediment delivery are first controlled, eliminating the need for re-dredging. Because dredging requires the disturbance of the lake bottom, any entrained toxic materials in the bottom sediment can be re-suspended and made available to the biological community in the lake.

Acidification

Atmospheric pollution by sulfur dioxide and nitrous oxides is among the principle causes of acidification of aquatic systems, through precipitation (acid rain) and to a larger extent through leaching from affected land. Other sources of acid deposits are industrial effluents and mining wastes. Concern about acid deposition has existed since the mid-nineteenth century, but the actual effects in Northern Europe and North America have been documented only since mid-twentieth century.

Most surface water is naturally acidic with a pH slighly less than 7. Adverse affects on the most sensitve aquatic species, for example the larvae of the salmonid fishes and freshwater crustaceans, start to appear when the acidity reaches a pH between 6.5 and 5.5. As the acidity increases, more tolerant species become affected with most aquatic organisms, for example, the blue-green algae, not surviving at a pH below 4.0. The affects of increasing acidity on fish include changes in the ion content of body fluids, leaching of salt from body fluids, clogging of gills, and reproduction failure. Shifts in lake biomass and reduction in species diversity have been observed.

Thousands of small Scandinavian lakes in particular (where neutralizing components are not present) have been harmed by the deposition of airborne acidified particles. In southern Norway, acidification has caused more than a third of the lakes to become fishless (Muniz 1984). It has been observed that increased acidity was accompanied by dramatic increase in dissolved aluminum levels (Dickson 1980). The toxicity of aluminum manifests itself in fish long before stress from the low pH (Muniz and Leivestad 1980).

To neutralize the acidity in these waters, lime can be applied. The active ingredient is calcium carbonate. Liming has been used extensively, for instance in Sweden, over a long period, and has shown reasonably good results., both in the survival of fish species and the reintroduction of others. Applied to surface waters, limestone dissolves slowly, resulting in a gradual increase in pH value. However, liming is expensive and must be repeated, and can "only be regarded as a short-term stop-gap procedure" (Swedish Ministry of Agriculture 1982). The need for treatment at its source is evident, but this requires international agreements and enforcement.

Fish depletion

Fish provide a large percent of animal protein consumed by the world population. In tropical developing countries, 60 percent of the people depend on fish for 40 percent or more of their protein (WRI 1994). The majority of the world's landed fish catch (87 percent) comes from marine areas. Of this, the annual freshwater catch grew 85 percent between the late 1970s and the late 1980s, while the annual marine catch increased by 35 percent. Recent catch statistics show that the annual marine catch may have reached a limit near 100 million metric tons and more pressure on freshwater fisheries may result. This could mean that lakes and reservoirs as sources of fish protein will become more important.

Most large lakes and reservoirs supply fish protein for human sustenance on either a commercial or recreational basis. The large lakes of Africa, including Victoria, Tanganyika, and Malawi may provide a majority of the protein needs of the resident population. Sustainable fish production requires information about biological and chemical variables in the water. The factors that regulate fish yield and fish growth rates are food availability, dissolved oxygen, temperatures at various levels of the water, and the seasonal competition within and between species for food and spawning grounds. Fish production is indirectly related to nutrient concentrations.

In eutrophic conditions, either natural or induced by heavy inflows of sewerage or agricultural run-off, a high biomass is likely to be found. A positive correlation between fish yields and total production of algae has been reported in some cases, and in other cases there is a positive correlation between fish production and bottom-dwelling invertebrates (Hecky et al. 1981). The same increase in nutrients may promote intense algal blooms, however, leading to increased biological oxygen demand, oxygen depletion in bottom waters, and a change in the fishery.

New developments in watersheds, such as dams and reservoirs to control the annual distribution of water, frequently cause large losses in floodplain fertility and species diversity. Floodplains are needed by many species of fish for reproduction and for refuge. The natural, seasonal flooding, and drying are signals used by some river-dwelling animals to begin to reproduce or migrate. Proper timing of water level drawdowns in upstream reservoirs is therefore important, expecially to enhance spawning of certain species. Reservoirs with the primary objective of drinking water supply, irrigation water, and power production, can include fishery production (see generally Gasschignard and Berly 1987).

Where non-native exotic fish have been introduced, intentionally or by accident, changes in biomass, species composition, and production may occur. Although total fish production might remain the same or even increase, the change in species composition might disrupt the livelihood of local human populations, and can affect the biodiversity in the lake. An example of effects of this kind is the introduction of the species perjury into Lake Titicaca, where this very hardy, pollution-tolerant fish appears to be thriving as the lake waters are becoming more eutrophic and polluted with human wastes and pathogens. Indigenous people, with little choice, are exposed to great health risks from eating the contaminated fish.

The case of Lake Victoria is even more egregious. With the introduction of the Nile Perch in Uganda in the late 1950s as an experiment in economic development, over 200 species of fish that were endemic to this lake are now thought to be extinct. The loss of these smaller fishes, which contributed substantially to the diet of the local indigenous population, has forced these people to seek other non-fish protein sources, such as maize. They cannot afford to buy the Nile Perch because it is too expensive. Some have resorted to eating rotting perch carcasses from fish processing plants, increasing the risk of disease.

Excessive water diversion

Water diversion or its disruption probably has the greatest effect on the over-all quality and ecological health of a lake or reservoir ecosystem. Diversions of water out of a lake system, frequently tied to non-sustainable economic development schemes, are used for supplying irrigation water, drinking water, or industrial process water to other regions that lack water. Dramatic examples of the adverse effect of diversion include the Aral Sea (Annex 9), Mono Lake (Annex 3), and in the Lake Chad basin (Annex 7). In the case of the Aral Sea, more than 60 percent of the incoming flow to the lake was diverted for irrigating cotton and other agricultural crops (Precoda 1991). Consequently, the surface area of the lake was reduced by 55 percent to 37,000 square kilometers (km^2), its volume was reduced from 1,090 km^3 to 340 km^3, and its salinity has almost tripled (Annex 9).

Diversion has other consequences. Raising and lowering of lake levels destroys marginal habitat and wetland areas that are necessary for nursery grounds for terrestrial and aquatic species, buffer zones to erosive wave action, water recharge, and nutrient sinks. Lake lowering increases difficulty in water access, disrupts water intakes, erodes embankments, and disturbs recreational use.

Downstream ecological degradation can occur when reservoirs are not operated with a view of protecting environmental quality. Withholding water for irrigation, flood control, power production, and other economic development purposes can have and is having severe ecological consequences. A sufficient and sustained supply of water is required to maintain marginal habitat, nursery areas, and wetlands. It is reported, for example, that wetland habitats below the Kariba Dam in Zambia and Kainji Dam in Nigeria have been drastically reduced. The Inland Delta of the Niger River, the Lake Chad Basin, the Sudd, and the Kafue Flats in Africa are similarly threatened. Dams prevent downstream regular flooding required for recessional agriculture and for replenishment of floodplain and delta lakes that are critical nursery areas for riverine fisheries. Artificial floods should be created from reservoir releases during normal high flow season to sustain the downstream floodplain zones. Diversion can also cause problems in receiving waters as a consequence of higher water levels, such as in the case of Southern Indian Lake in Manitoba, Canada (Hecky et al. 1991).

5
International Lake and Reservoir Basin Issues

Across the world, transboundary water pollution and inappropriate water management practices interfere with society's use of lake water. While almost 300 treaties exist for transboundary water management, most of them deal with the quantity aspect of water use by sectors and do not concern themselves with pollution or ecosystem issues. The large African lakes, the Great Lakes of North America, small lakes in the deltas of international river basins of Africa, and lakes like Titicaca, Biakal, Constance, and Geneva all have drainage basins covering more than one nation.

In some cases, countries have taken action to reduce pollutant loading. Lake Vanern, Constance, Geneva, and the five North American Great Lakes are good examples of this (see Annexes 10 and 11). Pollution from man-made toxic chemicals is particularly difficult to address not only because the source of pollution must be reduced but accumulation of persistent toxics in the environment (contaminated soil, groundwater, lake or river bottom sediments) must be investigated and expensive remedial actions undertaken. Many areas of the Great Lakes have highly contaminated bottom sediments. This is true particularly in harbors and bays containing industrial sites, and in river deltas where upstream industry has discharged wastes containing toxics. The Great Lakes experience (Box 5) offers an example of international cooperation by two nations involving all stakeholders in the process of identifying the extent of the problem, understanding the health threats to the ecological system and the people of the basin, and defining appropriate mitigation, including the institutional and technological requirements.

It has been recognized that stopping pollution discharges alone is insufficient. Accumulated deposits must be treated as well. The remedial action plans (RAPs) being utilized in the Great Lakes basin for 43 "hot spot" watersheds (Annex 11) represent good examples of the enormous level of effort needed to restore lake ecosystems once discharge of industrial toxic wastes has occurred. Even radioactive particles (radionuclides) can be a problem in international lakes as is shown by: a) uranium tailings runoff at Elliott Lake, Canada, draining to Lake Huron, b) fallout from the Chernobyl disaster that was deposited in Belarus and is now leaching through soil to contaminate Kiev's water supply reservoir downstream, and c) radionuclides released to a river in Mongolia that are degrading the quality of Lake Baikal in Russia.

Even simpler, less sophisticated pollutants as well as poor management practices can cause international problems. Sediment from deforestation in Malawi, Tanzania and Mozambique inhibits fish reproduction in Lake Malawi. In Lake Titicaca, eutrophication stimulates algae which shade out aquatic plants that are essential to the ecosystem and to indigenous peoples. In Lake Victoria, the combined effects of overfishing, introduction of nonindigenous fish and aquatic weeds, and sewage pollution causes very complex degradation of the lake and problems for residents of the three riparian countries. In the case of lakes connected to oceans, accidental introductions of nonindigenous species in ballast water from ships causes extreme damage. The Lake Erie aquatic ecosystem, which provides $200,000,000 in fishing benefits each year may soon collapse because of the Zebra Mussels introduced in ballast water of ships from Europe. Billions of dollars need to be spent to prevent fouling of intake and discharge pipes clogged by the mussels.

Poor water management in terms of upstream diversions or changes in flow regimes downstream of reservoirs also degrade lake ecosystems, causing economic damage, and human suffering. In the case of delta lakes in international river systems (such as the Ganges delta) that are essential for fish propagation, upstream withdrawals of water cause salinity to increase and ruin the functioning of the lakes. Impoundments for hydropower also prevent annual floods that are essential for floodplain lakes (and for fish propagation and nurseries), flood recession agriculture, and recharge of groundwater supplies. This destroys the fishery of African rivers, displaces floodplain peoples, and dries up their groundwater supplies. Proper water management practices are needed to ensure that there is sufficient water left in international rivers to recharge delta lakes and that reservoir releases include the provision of an annual downstream flood to enable floodplain lake ecosystems and floodplain agricultural communities to survive. Disruption of sediment flow to coastal zones by dam construction can cause coastal erosion, and in West Africa entire coastal towns have been washed away.

Two other international issues also raise problems for lakes. One involves a dynamic link between lakes and groundwater supplies and the other involves atmospheric deposition to lakes. Lake Chad (Annex 7) represents a good case of links between lakes and underlying groundwater drinking supplies. As more water has been diverted for irrigation from the lake basin (and as climate became drier) the lake has shrunk, wetlands that sustained the lake fishery have dried up, and groundwater levels have declined. The international groundwater around this lake is connected to the lake and depends on lake inflow to sustain its level. Too much diversion for agriculture also means that groundwater sources in this dry climate may not be sustained.

Toxic rain (deposition of toxic chemicals in wet and dry forms) is disrupting the ecosystems, bioaccumulating in fish, and posing health hazards to humans. As Annex 1 notes, mercury deposition is causing ecosystem contamination and health risks to those eating fish in northern temperate zones. Releases of other toxic substances contaminate the air, and are then deposited in lakes and their watersheds. A good example is toxaphene (and now fresh DDT) showing up in the North American Great Lakes. DDT being applied in Latin America can be volatilized and transported long distances to be deposited in Lake Superior in rainfall and in dry form (Schmidtke 1988). Clearly, airshed considerations and deposition of toxics must be addressed to protect sensitive lakes. This is an international, transboundary issue. Priority areas are North America,

Europe, and Asia, especially China and India. Energy conservation measures (demand management) represent the most cost-effective solutions. An example of managing an international lake basin is presented in Box 5.

Box 5. Institutional Experiences in Managing the International Great Lakes Basin
The framework for cooperation in the Great Lakes Basin is contained in the Boundary Waters Treaty signed by Canada and the United States in 1909 and subsequently amended by specific water quality accords adopted by the two countries in 1972 and 1978. The Great Lakes experience has successfully identified a number of factors that should be considered in designing any effective transboundary, multi-jurisdictional water resources management process.

A written agreement, signed by all affected jurisdictions was found necessary where the basic tenets or purpose for the need to cooperate was identified. The agreement commited the two governments to cooperate in the pursuit of mutually agreeable goals, in this case protection of the quantity and quality of the water in the Great Lakes. The agreement provided the vision for the participants, presented the general structure for cooperative interaction, and established the basic ground rules on operations. Time frames for accomplishment were subsequently identified and agreed to by the parties as specific accords were developed. A schedule of progress reports to the stakeholders and parties outside the agreement was also found necessary and helpful as a means of ensuring progress.

The Water Quality agreement of 1978 focuses efforts on pollution prevention rather than end-of-pipe treatment, as it was found cheaper to prevent pollution, particularly toxic pollution, than to clean-up or restore a site after it has been contaminated. The process encouraged concepts such as zero discharge or virtual elimination of toxics pollution coupled with specific commitments for cross-sectoral, multi-media pollution prevention initiatives. This approach emphasizes the introduction of management principles such as source reduction, process redesign, recycling, and reuse.

This multi-jurisdictional approach to water resources management has attempted to include all of the major stakeholders who are or will be affected by any resulting policy or specified actions. Through this approach, each of the participants has gained a sense whether their interests are getting a fair hearing, and the results of negotiations reflect their participation. This assurance established buy-in by all concerned parties at the earliest stages, and has allowed implementation of agreed to approaches to proceed more quickly. Effective, not token, public participation, including that of NGOs, is essential in this process. Every effort has been made to include them in the multiple aspects of project design, assessment, and remediation strategies.

The Great Lakes process is credible and transparent; there are sufficient checks and balances in the system. The process includes independent scientific and policy "advisory boards" that provide unbiased information, analysis, and recommendations to the responsible body, the International Joint Commission established by the original 1909 treaty. The Commissioners are high-level officials appointed by the government of each country. They report findings and recommendations, and negotiate implementation strategies at the senior levels in the governments. This helps to assure that the work of the international body will get the necessary backing of the governments. A well-staffed and supported impartial secretariat is also part of the process.

The Great Lakes experience shows that joint institutions can be established among nations to improve the management of transboundary lake and reservoir basins. Processes can be initiated through these joint institutions to achieve environmentally sustainable development objectives of mutual interest and concern. These institutions can have built-in flexibility to changing focus over the years and can adjust to transient environmental and external institutional conditions.

Source: LeMarquand (1993).

6
Economics, Regulations, and Institutions

Lakes are considered in natural resources and environmental literature to be a common property resource, accessible to averyone. Therefore, many of the analyses applied to any common property resource may be appropriate to lake water extraction and pollution problems. This section includes a general review of main approaches to pollution control, drawn on Baumol and Oates (1992) and Downing (1984), and a review of several case studies that include economic analysis.

General economic principles

A basic behavioral assumption in economic theory is that individuals or parties in a given economy act to maximize their own self interest, subject to a set of constraints imposed by available budget, technology and governmental policy. (Self interest--dictated by what economists define as utility functions--depends on specific circumstances.) This basic assumption, involving many parties, may lead to efficiency in resource allocation. This efficiency is achieved through the market mechanism. If, in that market, one individual's decision does not affect other individuals' outcome, that is, each party's decision is independent, then the market is called perfectly competitive. In a perfectly competitive market that economists call "perfect competition", the self interest of the parties leads to an efficient outcome for the whole economy.

The perfectly competitive market model is not fully applicable in the case where one party's decision affects another party's outcome, such as in the case of pollution. This is called the externality effect of pollution (The discussion here will not expand on the theory of externalities.) The reason is that if pollution is generated by one party, it imposes costs on other parties, costs that the polluting party does not bear. If the polluting party is a producer, it does not include the pollution-related cost in the supply curve of the good it produces and thus, the market mechanism collapses. This introduces inefficiency into the system and results in what economists call "market failure." If no particular party is directly affected by the pollution, then the externality needs to be born by the society, which pays what economists call "social cost" associated with the pollution (either the mitigation cost of the damage to environmental amenities, foreclosed opportunities, or any human health problems).

In the case of lakes, the externalities might be multidirectional. This is particularly true for lakes shared by more than one party (e.g., country), and for lakes used for more than one purpose. For example, an industry that uses a lake's water for cooling and as a means for pollution treatment (dilution), may affect the water quality not only for recreation, fishing, drinking, and other uses, but also the quality of water for cooling (by enhancing corrosion due to elevated levels of chemicals in the polluted water).

The aforementioned market inefficiency can be corrected in various ways. One way is through regulation imposed by a governmental authority. Several things can be regulated by that authority. In the case of lakes, the regulatory variable can be the total amount of emissions (of pollution) into the lake. If the correct emission quantity is imposed and allocated in the right way among the different polluters, then efficiency, or near efficiency (because of the additional monitoring and enforcement costs) can be achieved. One can see immediately, however, the challenge of how to allocate the total amount of regulated emissions to the lake among the various polluters. This will be elaborated in the next section.

In order to apply market principles to pollution control, economic theory suggests placing a price on pollution emissions. In this way, pollution becomes more expensive, and indirectly is controlled by the market mechanism. The relative efficiency of this mechanism compared with the regulatory approach is, however, questionable, particularly in cases where the pollution is not linked to a specific source, such an industry discharge. (Economists define this situation as a "nonpoint source pollution". The discussion here will not develop this concept. (See Russell and Shogren (1993) for a more detailed analysis and for examples.)

Some economists argue that environmental quality is a public good, so affected parties avoid taking action. Therefore, government is called to take control on behalf of society. In this case, the political power of interest groups may affect governmental policies so that the desired efficiency is not achieved. Lobbying by both polluters and parties affected by the pollution is very common as a means to affect the resulting policy, and the outcome will depend on the relative strength of the lobbying efforts of the parties (for a more technical analysis in the case of irrigation water pollution see Loehman and Dinar 1994).

Pollution control

Governments can control pollution in various ways. This section provides a non-technical summary description of these interventions with emphasis on lake water pollution. (See Baumol and Oates (1992) and to Downing (1984) for a more detailed discussion.)

Government control of lake pollution via regulation may include in-lake "water quality standards," "effluent standards," or combinations of both. Each regulatory approach includes monitoring and enforcement costs and contains inefficiencies due to the common resource nature of lakes that has been explained previously. Another pollution control approach is the "incentive approach". Incentives to reduce pollution can be in the form of placing a tax (fee) on effluent volumes (Pigouvian taxes), or providing subsidies to the polluter to adopt pollution reduction technologies. The desired rate of subsidy or tax on pollution should be such that it will bring the polluters to an optimum point (private optimum) on their production-pollution curve that will coincide with the social optimum. In other words, the incentive internalizes the cost of the pollution damage to society within the polluters' business decision, and thus eliminates or reduces it in an efficient way.

In the case where there are many pollution sources with no explicit identification, and where no exact measure of each source can be made, effluent taxes are not effective in solving the problem. In this case, known as nonpoint source pollution, taxes on specific components of production (inputs) are one of the preferred incentive measures. Such taxes may be easier to monitor, and they can have a similar effect to that of Pigouvian taxes if the relationship between input levels and pollution are known. The taxes are imposed on inputs directly associated with the pollution. For example, if a lake is polluted by pesticides, or fertilizer residuals from agricultural activity in the lakes watershed or along its shoreline, the amount of pesticide or fertilizer used by the producers should be taxed. An appropriate taxation may cause a reduction in the use of these inputs by the farmers, and thus reduce the amount of pollution. Sometimes, the effect of the polluting input is enhanced by another input. As an example, water serves as a carrier of chemicals through the soil. Improper use of water for irrigation, for instance too much, may result in increased amounts of pollution in the form of dissolved chemicals carried to a stream or lake. In such cases the incentive measure might be to tax both inputs, the water used as well as the chemicals applied. Unlike other inputs, however, the volume of water is not always easy to measure because of certain delivery setups. This reduces the effectiveness of the polluting-input tax approach.

In cases where management practices rather than input levels may reduce (or increase) pollution, the regulatory agency might consider incentives in the form of subsidies (or taxes) for the application of improved technologies. The additional societal cost associated with lost public funds that are spent as subsidies, can be justified by the reduced cost to society of correcting environmental damages. For a more technical analysis of the economic aspect in the case of pollution-reducing irrigation technology selection see Dinar and Zilberman (1991).

Another approach to pollution control involves pollution "trading". The principle behind this approach requires governments to calculate the sustainable amount of a given type of pollution that an environmental unit, for example a lake or river, can assimilate in a given period. The polluters are then allowed to purchase pollution allotments which are identified in their permits. The proceeds from the permits are used to operate the trading program, including the monitoring and enforcement of the pollution quotas. Assuming that all pollution sources in a given permit market have the same effect on the environment, polluters could trade pollution permits among themselves. This approach has been implemented in industrial regions around the world (Atkinson and Tietenberg 1991; Tietenberg 1988 p. 419).

Approaches to environmental damage estimates
Literature on environmental damage demonstrates the complex nature of determining quantitative damage and cost estimates of environmental pollution, and in particular, of lake contamination. It is not difficult to estimate the costs associated with lake restoration, or with direct measurable damages to lake services (such as lost quantity of water for hydropower, or water sold to agriculture) due to water quantity and quality reductions. A more complicated task is the estimation of losses from damages that affect enjoyment from the environmental amenities that do not have a market value.

If markets exist for environmental goods and services, their economic value would be the sum of the actual expenditures to purchase them plus the associated consumer surplus. The transactions in the market mean that ownership of the commodity changes. When ownership and pricing are not well determined, special techniques are needed to estimate consumer demand for these commodities and services on a comparable level to the "conventional" commodities.

A seminal paper by Vickrey (1992)[1] that addresses the limitations and possibilities of a market mechanism approach to air pollution control is very relevant in the lake pollution context as well. The paper summarizes the advantages that the market method has over other methods of control such as subsidizing pollution prevention technologies, taxes, and quotas on pollution abatement. Examples where market solutions are superior and inferior to other methods were provided.

The environmental economics literature on water quality, and other environmental amenities has developed several tools for this purpose. They include (a) models of hedonic prices that take into account quality aspects in the demand for marketable goods and services; (b) travel cost models that measure the willingness of individuals to pay for travel to certain recreational and environmental cites; (c) household production functions that include components that substitute for environmental attributes; and (d) experimental methods using hypothetical settings (contingent valuation method), that were designed to provide answers to questions about the value associated with damages to environmental amenities that do no have market values. The interested reader can find a more detailed explanation in Braden and Kolstad (1991), and Smith (1993).

Results of a contingent valuation method application to evaluate lake water quality problems (Loomis 1987) is presented for Mono Lake (Annex 3). However, the special nature of lakes, which have several uses and many users makes the evaluation of damage to the lake's water quantity and quality much more complicated. An example of such complications is presented in McConnell (1990). McConnell demonstrates that if pollution of a lake causes economic damage

---

[1]The original paper was presented at the Air Pollution Control Association Meeting, Cleveland, June 1967.

because the lake is less attractive to two or more activities--say recreational fishing and swimming--then an investigation to estimate separately each of these amenities may lead to an overestimation of the value associated with the improvement in the lake's water quality. This leads us to several examples to demonstrate the nature of recent empirical work on lake pollution.

Economic applications in selected case studies

A number of case studies provide an appreciation of a variety of economic models and techniques that are available and can be adapted and applied to lake and reservoir management.

LeMarquand (1989) discusses institutions for sustained economic river and lake basin development with emphasis on developing countries, particularly in Africa. The experience in Africa suggests that the real challenge to achieve environmentally self-sustained water resources development is associated with the integration of physical development with the social and environmental outcomes. The river-lake basin development concept includes: multipurpose projects; basin-wide planning and management; basin institutions; and regional development. The Tennessee Valley Authority is used as a basic model for integrated planning that is then modified for river-lake basin development in developing countries with the additional analysis of multi-donor financing. One main lesson to be learned from this study may be captured in the following citation: "A planner who looks only to his country's section of the basin will find some development possibilities. But generally the net benefits from his plans and the plans of the other basin states summed will yield total net benefits less than what might have been achieved by planning as if boundary lines did not exist" (p. 136).

Becker and Easter (1995) analyze the dependency among eight US states, two Canadian provinces, and the two - United States and Canada - federal governments in the Great Lakes region. Their study estimates the economic outcome of different water diversion policies for two main uses--hydropower and navigation. The study found that because the big users of water are downstream, large economic losses can result if downstream water rights are ignored. However, while diversions of water may be economically desirable, care has to be taken through proper regulation to assure downstream needs are met while minimizing large environmental and economic losses upstream. Therefore, non-cooperation in water diversions may offer big gains to some parties, but cause bigger losses to the basin as a whole. This is known in the common literature as "the tragedy of the commons," where parties act to minimize losses and not to maximize profits.

Stark et al. (1995) introduced a procedure for management of water in lakes and reservoirs by local interest groups, as opposed to centralized dictated policies by national level government agencies. The approach of looking at a water resource from its characteristics viewpoint may offer an alternative to price-variable forms of analysis. This means that water is looked at as a vector of "water components",and not as one product. Having several components, water can satisfy different users needing the water for various purposes. The procedure, called Management by Interest Group Analysis (MIGA), has several steps: (a) identification of interest groups and the water resource characteristics; (b) collection of value response data; (c) estimation of group welfare functions; (d) comparison of group welfare functions; (e) checking Pareto optimality conditions; and (f) estimating political power measures. The approach has been applied to Lake Shelbyville, Illinois. The lake serves several interest groups: upstream agricultural producers, camping, fishing, swimming and boating recreationists, down-stream flood control beneficiaries, lake shore property owners, and a far down stream group that receives navigation benefits. Results from the case study suggest that the current lake management policy is sub-optimal and that it may be possibile to increase welfare for some of the parties involved without changing the other interest groups' welfare.

Hitzhusen et al. (1995) address the sedimentation of lakes by agricultural soil erosion. Dredging and sediment handling in lakes is associated with high cost. They develop an institutional context for the dredging of sediments from Ohio's state lake parks in an attempt to rationalize economically the dredging activity. The approach includes public choice models, based on physical, political, and economic variables, for the dredging decisions, and a hedonic pricing model that estimates sedimentation effects on lakes. The results suggest that lake property owners and boat recreationists could bear a higher proportion of the present and future dredging costs through property taxes and higher fees, respectively. The approach also suggests that better coordination and control could lead to more efficient solutions than dredging alone. This also means the assumption of soil erosion damage costs to the lake by the upstream farmers.

Cann (1989) describes the restoration effort of Lake Hopatcong in New Jersey. The 2,700 acre lake is located in an accessible location to the residents of New Jersey. It is bisected by two counties and four municipalities The increasing number of visitors and recreational users with no adequate pollution control measure began to put stress on the lake and its surrounding watershed. Water quality declined as lake cottages, originally designed for summer usage, were converted to year-round residences. As the lake showed signs of increased weed and algae growth (in 1959), private owners of property at the lake began spraying herbicides to kill the weeds. As the water quality began to worsen, a regional planning board for the lake was established (in 1962) in order to restore the lake's water quality for the benefits of the neighboring population. Although progress in the first years was slow, the joint effort of all interested parties gained momentum. In 1982 the Lake's Regional Planning Board was awarded a US EPA grant to understand the pollution patterns in the lake. The results of that study helped the Board in its effort to regulate septic system drainage into the lake, control construction in the watershed, and develop other regulations to control pollution later adopted by the municipalities surrounding the lake. The most successful outcome of this project is the enhanced cooperation among the individual property owners around the lake who have supported the Board and decreased herbicide spraying.

Conclusions from the above analyses that include also several applications (See additional applications in Annex 13) suggest that a number of issues should be considered when evaluating the costs and benefits from protecting and restoring lake resources:

1. Lakes are complicated ecosystems that are very sensitive to human interventions.
2. Externalities exist in several levels: Between lake ecosystem components, between lake ecosystems and human intervention (either development or restoration activities), and between users of lake beneficial uses.
3. The links between the ecosystem and the development activity (such as a dam) is dynamic. The lake ecosystem may assimilate negative externalities but the extent is limited. In certain cases ecosystems can recover, but in many other cases the damage is irreversible.
4. The institutions for a successful management of lake systems should include all stakeholders. These stakeholders should also be involved in all levels of discussions on development, on regulation, or on restoration of the lake.

7
A Comprehensive Approach to Managing, Protecting and Rehabilitating Lakes and Reservoirs

A comprehensive approach to water resources management is needed that is cross sectoral, that integrates ecological and development needs, and that is based on holistic considerations of the carrying capacity of the water environment. The World Bank (1993) recently adopted such an approach in its water resources management policy, which reflects several years of consultations within the Bank as well as with borrowing countries and NGOs.

World Bank water resources management policy

In this comprehensive approach, the watershed, or the hydrological basin are the management units upon which changes are evaluated in how sectoral development activities are to be carried out and where priority environmental interventions are required.
sustainable use of lake and reservoir ecosystems require special programs, special priority, and special expertise for proper management. This expertise needs to incorporate the complex relations that govern water quality within a hydrologic region, as well as conservation of living resources (fish populations), and preservation of habitats such as wetlands. Actions to combat existing problems and prevent future transboundary degradation must focus on changing the way human activities in different sectors are carried out. In doing so the capacity of the particular water environment to sustainably support that activity is not exceeded.

In many instances, this capacity to support sectoral activities has already been exceeded, and action programs to restore the functioning of ecosystems or to remedy human risks are needed urgently. Interventions will be different in different regions as part of operationalizing sustainable development policies with different sectors and different government ministries.

The Bank policy stresses that stakeholder participation, legal and regulatory reforms, decentralization of water service delivery, proper pricing and economic policies, and an emphasis on environmental protection and restoration are all integral to this comprehensive framework for improving water resources management. Self-financing mechanisms are based on who receives benefits from exploiting lake ecosystems and are the key to supporting the sustaining management institutions.

The World Bank water resources management policy represents a new commitment to support countries in implementing this comprehensive approach. It signals that the highest levels of government should recognize that water resources and watersheds must be managed as valuable natural resources to meet multiple uses rather than just inputs to specific sectoral activities. It also seeks to balance the need for holistic, ecosystem-based approaches for sustainable management of water resources and their drainage basins with the advantages of relying on markets, pricing, and economic policy reforms to improve water management. Current practices are simply not sustainable from either economic or environmental perspectives.

The Bank will provide assistance to member countries to develop a comprehensive approach to water resources management suitable for the country's needs, resources, and capabilities. The Bank recommends (Le Moigne et al. 1994) a two-step process. First, a cross sectoral water resources assessment should be prepared by an inter-ministerial team of country officials. Then, a water resources management strategy should be formulated to address priorities in each river or hydrological basin--including environmental priorities. Lessons learned about degradation of lake and reservoir basins show that as countries undertake this two-step process, lake and reservoir basins should be accorded priority so that benefits they provide for sustaining the ecosystem and human society can be maintained.

Resolving conflicts

In the absence of property rights and markets for lake waters, conflicts over competing uses are unavoidable. There are several layers of potential conflict that can be observed. Conflicts may involve some or all layers. First, there is a potential conflict between development and sustainability of lake basins. Second, there is a conflict between uses and users of the lake beneficial services. This conflict may include users within jurisdictions, or sometimes conflicts between nations.

An environmental policy can be implemented by enforcement or by consent of the parties involved. Enforcement of the policy is necessary in cases where there are opposing interests of the parties involved, which may lead to disregard or violate the policy. Consent will be the result of a process in which the parties involved will realize some joint benefits from the proposed policy. With the traditional regulation-tax-subsidy scheme, there does not seem to be any need for cooperation, and conflict of interests does not matter.

Because lakes are a common property resource with externalities in their use, lake utilization can not always be analyzed using a market mechanism. The literature on common property resources provides general examples that support the failure of the market system (Randall 1981). To model such environmental problems, game theory is an appropriate approach that captures and incorporates all economic and political aspects (See Box 6).

For example, to cooperatively solve the problem of lake acidification, as well as other related lake water pollution problems, the parties involved should realize some mutual benefit(s) that can be achieved only through cooperation. In cases of cooperation, each party needs to participate voluntarily, and to accept the joint outcome of the cooperative process. Once a joint interest exists, however, a major problem remaining to be solved is the allocation of the associated joint costs or benefits.

Traditionally, engineering solutions have been used as a tool in managing basin-wide water-related problems. Economic approaches have been less common. Penalties, taxes, and other incentives intended to regulate should be introduced more widely, especially in

Box 6. Conflict Between Private and Social Interests in the Coal Industry

Solutions to environmental pollution that are preferred by private individuals may ignore many aspects involving social and externality effects. Therefore, comprehensive solutions, such that the ecosystem approach, to lake pollution may prefer prevention over mitigation. As an introductory note it is most interesting to cite the following discussion from Wooten (as reported in White 1985). The paper provides the Coal industry perspective on liming acidified waters. While recognizing that acidification is caused also by acid mine drainage, the coal industry prefers mitigation rather than prevention measures. To demonstrate its support in the liming technique, it is being argued that "[P]erhaps one of the most important considerations in the use of liming as a mitigation technique is its high degree of cost effectiveness in relation to controlling acid deposition precursors. To install a flue gas desulfurization system to a 500 MW steam electric generating unit using medium- to high- content sulfur coal would require a capital expenditure in excess of $100 million. To own and operate this system over the 30-year life of the plant would require an annual revenue of approximately $41 million. It has previously been reported that Sweden will spend an almost identical sum of money, $40 million, in 1986 to lime approximately 20,000 bodies of water. It takes little imagination to calculate the cost effectiveness of liming 20,000 water bodies as opposed to reducing the emissions from one major power plant which, based on the current state of the knowledge, may or may not affect the acidification process."

Why should the coal industry agree to accept the privately expensive technology based solution while there is the liming solution? It is true that the liming solution might be socially less desirable, but it is clear that from the coal industry point of view, it is most preferable since it allows the industry to continue with the same practices with no additional costs, regulations or interventions.

Source: White (1985).

developing countries. Corrective measures applied to pollution sources in lake basins are frequently difficult to relate in economic terms to a directly measured improvement in the condition of a lake since improvements are progressive over time and some times marginal. Much better knowledge is needed to evaluate how economic mechanisms can help the process of integrated lake management.

Institutional components in multidimensional lake management are a crucial issue. Overall basin-wide institutions should be developed, taking into account the existing conflicting objectives. These institutions should be flexible so they can adapt as problems arise. An example of creating cooperative institutions for improving the environmental management of Lake Victoria's Basin is presented in Box 7.

Box 7. Lake Victoria Environmental Management Program

Lake Victoria is shared by Kenya, Tanzania and Uganda. Resource use by one riparian state affects the economic and environmental activities of the other riparians. Therefore, it has been recognized that resource development and management by the riparian governments within national jurisdictions has to proceed and be coordinated within a regional cooperative framework. This framework needs to be backed by a strong political commitment from the basin countries.

Recognizing the need, the Governments of Kenya, Tanzania and Uganda have agreed to develop a regional cooperative program for the management of Lake Victoria. Through a series of meetings (in Arusha on November 1993, and in Nairobi on May 1994) the countries reached understanding on objectives, priorities, and organizational arrangements for the program preparation. And finally, in Dar es Salaam, on August 4 1994 the countries signed a multilateral agreement on the environmental management of Lake Victoria and its resources.

The regional cooperation will focus on:

- Managing fisheries
- Controling water weeds (*Hycinth*)
- Monitoring water quality
- Managing pollution from agricultural (land use) runoff
- Managing pollution from urban areas
- Managing wetlands.

The countries agreed on lead responsibilities in preparing the regional program: Kenya will assume the lead for water quality and land use activities, Uganda will lead on fisheries management and water weed control, and Tanzania will host the regional secretariat serving the regional policy and steering committee.

Source: UNDP (1994).

8
Summary: Priority Issues and Next Steps

Apparently there is a serious decline in the overall quality of lake and reservoir resources world-wide. The importance of lakes and reservoirs to efficient and sustainable economic development of nations is well recognized. The lake-specific case study information contained in this paper, combined with limited available country-wide survey information points to an urgent need for countries to restore and protect their valuable lake resources. Many of these lakes are located in developing countries or industrializing countries that only now are recognizing the staggering economic and social impact that pollution, overfishing, and habitat degradation are having on their lakes. Once these natural systems are severely degraded they often cannot be fully restored; they can at best only be improved to a level where they can meet basic functions and society must bear the increased costs and risks to human health. Degradation therefore, means that people can be, and frequently are left with a residual penalty.

Experience reported in the reviewed case studies in this paper and elsewhere shows that polluted lake systems can be restored, at least partially. The paper presents a number of examples where this laborious, time consuming and very costly process has been undertaken successfully. Experience shows however, from an economic and environmental standpoint it seems prudent to prevent the problem from occurring or becoming worse, rather than turning to radical and costly restorative procedures. Preventing the problem also diminishes pollution externalities such as toxics contamination to humans and wildlife, the displacement of populations, the worsening of hunger, and the spread of disease which often afflicts poor and indigenous populations the hardest.

It appears that the international community has an outstanding opportunity to assist the developing countries through preparation of guidance and establishing financial support to advance more productive management approaches for these resources.

Priority issues

A great deal of guidance material on lake and reservoir management already exists for temperate lakes (for example, the publications of the UNEP/ILEC and the North American Lake Management Society). In order to capitalize on this information and to maximize the use of limited financial resources, these and other available materials should be evaluated and, where possible, modified to make them more useful for application in developing countries. Guidance for the management of tropical lake systems needs to be emphasized. As additional guidance is developed, the following points should be considered:

Comprehensive water resource assessments need to be conducted in order to allow countries to understand the condition of their lakes and reservoirs, so they can also provide the necessary means to establish important water resource management goals and objectives. These assessments, which are being encouraged by the World Bank and the UN System, should be pursued according to set protocols allowing intra- and inter country comparability, include inventories of pollution sources on a watershed basis, and identify areas of significant contamination and its impact on human health and the ecology.

Comprehensive watershed management programs that include meaningful water quality goals specific to the hydraulic characteristics of lakes and reservoirs need to be prepared. The differences between temperate and tropical lake situations need to be emphasized. In the design of management programs, every effort should be made to prevent the release of toxic substances to lakes and their watersheds. Introduction of exotic species should be prohibited, unless there has been adequate environmental assessment.

Recent approaches in sustainable economic development and natural resource management that incorporate "the precautionary" approach and the "polluter pays" principle should be included. In designing these programs, economic tools and sound financial polices should be applied in this goal setting process to make sure that the greatest opportunities for sustainable economic growth can be identified while assuring that the ecological integrity of lake basins is preserved.

As the priority action areas for protecting and restoring lakes and reservoirs become apparent through lake and watershed assessment, watershed control measures and best management practices need to be identified and developed for controlling specific pollution situations. These would include pollution from agriculture, and specifically irrigation return flows, silvaculture, mining, industrial pollution, urban runoff, and other contaminant sources, particularly for persistent toxic pollutants. Developing financial support to assure effective implementation of defined pollution control measures within watersheds and lake basins, should be given a high priority.

Water resource management programs for lakes and reservoirs should be broad in scope and watershed-wide in nature. Frequently, lake basins and their watersheds span a number of countries or within country jurisdictions. In this case, joint governance institutions, such as joint commissions, composed of high level government officials, should be established to improve management of transboundary lake and reservoir basins. Treaties and other legal instruments, granting authority for execution of decisions to the various states, and down to the local level, are essential for meaningful progress.

And, in keeping with policies for including stakeholder participation, countries should ensure effective public involvement in water resources planningfor lakes and reservoirs, with special consideration given to indigenous peoples and the poor.

Effective water resources management requires regular and consistent monitoring and evaluation. In the development of management strategies, relevant assessment measures for establishing progress will be required. These should include indices of physical and ecological parameters in addition to economic and public health considerations. The process will require transparency to assure technical adequacy as well as credibility among the stakeholders. A high priority should be given to efforts to incorporate this element in water resources management strategies for lakes and reservoirs and their watersheds.

Next steps

The World Bank and the UN System stand in a unique position They may not only foster an awareness of lake and reservoir issues among senior country officials, but also may provide tools for these officials already aware of lake and reservoir water quality problems. These officials may then take actions aimed at restoring and sustaining the multiple use of lakes and reservoirs. As part of the implementation of The World Bank's water resources management policy the policy, an opportunity exists to adopt strategies that place the proper emphasis on addressing the lake and reservoir issues. This is actually being done within the World Bank for its Africa Region as part of a water resources management strategy for Sub-Saharan Africa. However, other regions of the world also deserve attention.

In order to proceed in a most efficient manner, the Bank will use this background paper to shape a focused and comprehensive initiative to evaluate specific lake and reservoir issues on a global scale. The program scope will recognize the significant differences in the management needs of tropical and temperate lakes, and the Bank will use its portfolio of projects, combined with efforts of other international organizations, individual nations, multi-lateral efforts and academic resources to establish a clear understanding of these different management needs. The resultant issues analysis which will identify the status of the international lake resource base, the anthropogenic forces that are contributing to changes, the scope and rate of change in the quality of lake resources with an emphasis on ecological consequences, and various intervention strategies to mitigate the adverse situation. This process envisions the need for an international forum. This forum should include most notable lake and water resource experts, along with natural resource managers and economic development specialists from governments, industry, universities, NGOs, donors and other areas.

Out of this forum should come a consensus on the prioritized needs for lake and reservoir basin and watershed guidance that can be applied on an international scale. The outputs should focus on the needs of the developing countries, and establish clear management guidance that recognizes the differences between tropical and temperate climate lake situations.

Indeed, this issue is so complex and solutions are so complicated that a global commitment is needed among multilateral institutions, external support agencies, and developing country government officials. Such an international dialogue can help to focus attention on the sustainable development needs of lake and reservoir watersheds.

Evidence is overwhelmingly conclusive that the world's lake resources are in jeopardy. In many cases, the process of lake deterioration is advanced, threatening not only the lake but also populations depending on it. Investigative tools, technical and economic expertise, pollution prevention and abatement technologies, and institutions are available to address this resource management need. The action needed to restore and protect lakes should not be delayed. The scope, extent and dimension of the problem is such that it should be addressed in an international effort as well as national initiatives. Developing countries will require extensive financial and technical assistance to handle this mounting problem. The resolve, commitment and leadership of the international donor community and lending institutions is required now to address this serious issue.

Annex 1
An Ecosystem Approach in Lake Erie

The System
Lake Erie, one of the five Great Lakes in North America is approximately 388 km long and 92 km wide, with a mean depth of 19 m. The water covers an area of 25,657 km^2, and has a volume of 484 km^3. Over 270 km^2 of coastal marshes fringe western Lake Erie and its adjoining lake, Lake St. Claire. Up to 100 species of fish have been identified, and each year Ohio sport fisherman spend over 13 million hours catching 25 million fish from the lake. There are over 300 species of birds in the island archipelago in Lake Erie, and its wetlands produce more furbearers than the rest of the Great lakes combined. The main inflow to Lake Erie is the Detroit River which receives water from the upper Great Lakes of Huron, Superior and Michigan. The outflow is the Niagara River and the wetland diversion in Ontario, Canada (See Map of the Great Lakes in the Maps Section). Today, over 11 million people obtain their drinking water from Lake Erie.

Problems
During the late 1960s and 1970s Lake Erie was called the "Dead Sea of North America." Tributaries, such as the Cuyahoga River which flows through Cleveland, Ohio, were classified as fire hazards because of flammable substances floating on their surface. The human population around the lake increased from 3.8 million in 1910 to 11.2 million by 1960 (USPHS 1965). Domestic wastes were discharged into the lake and its tributaries directly rather than through the soils.

Increased industrialization, particularly as a result of World War II, resulted in larger volumes of industrial wastes entering the lake. To meet demands for food, more acreage was farmed using techniques that resulted in both higher applications of fertilizers and use of newly developed, longer-acting pesticides, many of which were washed into the lake.

Dramatic biological, chemical and physical changes occurred in the lake (Arnold 1969, Beeton 1969). The sport and commercial fisheries were severely affected by pollution and overfishing. The dominant species changed from blue pike, cisco, and whitefish to sheepshead, alewife, and carp. The increase in the number of the latter species reflected an alteration in quality of fish although the biomass remained the same. Nutrients and contaminants had reached high levels in Lake Erie during this period, and commercial fishing was no longer viable. The surface of the western and parts of the central basins became literally coated with algae (*Aphanizomenon flosaquae*). The decay of the biomass coupled with higher numbers of coliform bacteria prompted the closing of most recreational beaches. The decline in the quality of water, in terms of taste and odors for domestic consumption and industrial processing purposes, necessitated increased treatment and related costs.

A major cause of the biological and related economic and aesthetic changes was the increase of nutrient levels, particularly phosphorus. Between 1948 and 1962 the concentrations of soluble phosphate increased by a factor of five. The available nitrogen during the same period increased 30%. The reduced nitrogen/phosphorus (N/P) ratio (from 35 to 9.2), increased plankton productivity and affected water clarity.

A public outcry for improvement prompted the creation of a number of groups to deal with the ecological, and economic concerns in Lake Erie. The United States Federal Water Pollution Control Agency (later became part of the United States Environmental Protection Agency) was created to address ecological problems, while the Canadian Centre for Inland Waters was established to serve as a focal point for Canadian Great Lakes research and monitoring. The availability of new and increased funding initiated the growth of Lake Erie limnological programs. Regional and lake-wide surveys were undertaken, such as Project Hypo, a cooperative Canadian /United States program. Project Hypo demonstrated that oxygen depletion in Lake Erie was caused by excessive algal growth, and that eutrophication was a major problem of the lake.

The 1972 Great Lakes Water Quality Agreement between the United States and Canada, required all municipal discharges with flows in excess of 3,800 m^3/day to limit their effluent to 1mg/l of total phosphorus on an annual average basis. Programs to upgrade municipal sewage collection systems and treatment plants were initiated. Regulations to reduce the discharges of nutrients as well as toxic substances from industrial sources were imposed.

Agricultural programs, such as conservation tillage to reduce farm run-off and other practices to reduce non-point sources of pollutants have been introduced during the last ten years. Detergent phosphates were shown to be a serious concern connected with eutrophication in Lake Erie. The Great Lakes community turned to the legislative bodies of the United States and Canada to mandate change through phosphate detergent limitations in both countries, with which some states and the Province of Ontario implemented full or partial bans on phosphorus in detergents. However, the detergent phosphorus ban was not fully implemented until 1990. Phosphorus loading to Lake Erie has since significantly decreased.

Reversing the rate of eutrophication in Lake Erie has resulted in one problem being essentially corrected, but others remain. Combined sewer overflows result in dumping untreated sewage; degraded habitats limit fish and wildlife populations; and the presence of toxic substances and their effects on fish, wildlife, and humans.

The newest threat to Lake Erie is the invasion of the zebra mussel (*Dreissena polymorpha*), introduced in the 1980s through ship ballast water originating in Europe. This non-indigenous species is affecting the operational requirements of all industrial and municipal water processing plants. It is clogging all intake pipes exposed to Great Lakes waters and costs could run in the billions of dollars to retrofit these intakes with chlorinating treatment and maintenance procedures to prevent the further invasion of the zebra mussel. First observed in 1988, the mussels reached a density of over 700,000m^2 in 1992 at water intakes in the western basin of Lake Erie. At this density the mussels filter a volume of water 26 times as large as that in the western basin each day. This massive turnover of the water mass and the plankton has the potential to seriously disrupt energy flow from the plankton to small fish. Dramatic effects on the ecosystem have been observed. Food chain effects have been demonstrated. Research has shown that zebra mussels are beginning to alter contaminant cycling in the Great Lakes.

The governments of the United States and Canada are now directing their attention toward understanding the shifts in energy and contaminant transfer as entire functioning segments of the ecosystems are being replaced. The institutions created to address the earlier phosphorus and contaminant control measures are playing important roles in the management and in particular the coordination of the joint research programs in both countries, (including surveys and monitoring needed to determine data gaps), and in the outreach programs to inform the Great Lakes community of the severity of the threat these non-indigenous species have on the Great Lakes ecosystem.

Lessons Learned

The creation of strong institutions, either mandated by legislation or supported by other government and non-government processes, to identify and act upon ecosystem problems in Lake Erie, has provided an active, responsive network of policy and science to address and anticipate issues in the lake. The International Joint Commission, a bi-national organization is responsible for reviewing and evaluating the progress of governments in meeting the objectives of the Great Lakes Water Quality Agreement.

Ballast water guidelines, requesting all ocean going vessels to exchange their ballast water in mid-ocean prior to entering the Great Lakes have been developed voluntary and are designed to prevent invasion of additional species. This initiative is in cooperation with both the United States, Canada and the shipping industry. Collaborative efforts such as this to sustain policy initiatives work towards reducing risks of further effects on the environment. However, a mandated approach may be needed to ensure ship ballast water is exchanged. Such a regulatory approach would be viewed as a preventative measure and its execution would be less costly than to treat the organisms once they are established in the lake. However, the Great Lakes research community has resigned that the zebra mussels are in the Great Lakes and its tributaries to stay. Initial estimates indicate it will cost $2 billion in prevention and maintenance.

Alterations of the energy pathways in the Lake Erie ecosystem directly affects the dynamics of the fisheries resulting in losses to the commercial and recreational fishery of Lake Erie, and has imposed huge costs on industries and municipalities regarding the treatment of process waters at their water intake pipes. The situation of the Zebra Mussel clearly shows that environment and economy need to be linked to thoroughly understand the implications of human activities and their stresses on the limnological processes of Lake Erie. A multi-faceted, interconnected and sustainable ecosystem approach in decision-taking is necessary.

The research and monitoring developed to address Lake Erie concerns provides a model of information gathering that is multidisciplinary, interdisciplinary and transdisciplinary in nature. This type of investigative capacity is needed in identifying and anticipating ecosystem problems in lakes in general. The research information also provides a much needed data network to exchange with other information networks globally.

The approach applied to solve the Lake Erie pollution problem relied on regulatory actions. In the case of externality problems, such as in the Great Lakes, regulations and institutions are necessary for the success of a market-based operation in the long-run.

Annex 2
Pollution Loading, Fishing Pressure, and Introduced Species Endangering the Lake Victoria Basin Ecosystem

The System
With a surface area of 68,800 km^2 and a catchment area of 184,000 km^2, Lake Victoria is the world's second largest freshwater lake. It is a source of life and income for tens of millions of people inhabiting the area around the lake in Kenya, Tanzania, and Uganda. The lake basin supplies many beneficial uses. Major urban centers around the lake include Mwanza in Tanzania, Jinja and Entebe in Uganda, and Kisumu in Kenya, but there are hundreds of other settlements in the area. Fishing is an important part of the regional economy; the value of the landed catch of commercial fish for export is roughly $100 million with market values in the range of $250-300 million.

The importance of this large lake to the riparian nations' economies and to the well-being of the people cannot be understated. It provides reliable water supplies for industrial development, irrigation, livestock, and domestic purposes; serves as a convenient route for transportation; yields protein to feed the poor and fisheries to provide income; serves as a magnet for recreational activities and commercial development for tourism; and (as a shared international lake), it helps to generate a sense of common purpose and shared responsibility among the surrounding sovereign nations.

The Problem
Lake Victoria is experiencing an unprecedented mass extinction of fisheries and ecosystem degradation. The population growth of the riparian communities, which is among the highest in the world, creates many environmental problems. The lake's fishery is being overexploited and its waters polluted by industry, sewage, and agricultural activities. Reversal of this degradation will be more costly in the future and urgent actions are therefore needed now to prevent irreversible adverse effects.

The lake ecosystem has undergone substantial changes. The largest change has been caused by the introduction in the late 1950s of a non-native fish in Uganda--the fast-growing of Nile perch. This has resulted in enormous changes in the food web of the ecosystem. Degradation of the water quality in the coastal embodiments and shoreline areas has been caused by discharge of untreated municipal sewage water in the vicinity of the largest cities, as well as by the uncollected sewage from squatter settlements. Additional pollution discharged to the lake includes untreated or insufficiently treated industrial wastewater from textile mills, tanneries, vegetable oil factories, breweries, pulp and paper mills, and fish processing plants. Along with over-fishing and accelerated deforestation (which brings a great deal of mud to near shore waters), human activities are threatening a total collapse of the Lake Victoria ecosystem and a drastic reduction in its economic benefits and income for basin residents.

Eutrophication caused by excessive nutrients discharged from sewage and industry plants, agriculture runoff, and deforestation have caused huge blooms of algae, which deplete oxygen in lower depths of the lake. Several hundred small types of fish used to feed the algae, but these have now been eliminated by the Nile perch. The Nile perch are now forced to eat their young; the Nile perch fishery is declining, and investments in fish processing plants might be wasted if the overfished population of Nile perch crashes. In retrospect this experiment in economic development of a new fishery has been unsustainable. The loss of small fish also mean that people living in poverty have to subsist on protein-poor grains such as maize, instead of on protein-rich fish. The poor can not afford to pay the cost of purchasing the Nile perch (processed mainly for export) and they are furthermore subject to diseases transmitted by eating rotting perch carcasses discarded from fish-processing industries. While a few owners of fish processing industries have grown wealthy, the lake ecosystem has been virtually destroyed, species have been extirpated, poor people who depended on fish for food became more impoverished, and small-scale fishermen who invested in equipment will loose their investments since the current fishery is unsustainable. The water hyacinth is expanding lakewide, which threatens biological productivity of the lake and impedes lake transportation. The disease Bilharzia is increasing because the little fishes which used to consume the organism's host snails have themselves been eliminated by the voracious Nile perch.

Lessons Learned

The attempts of economic development in Lake Victoria basin placed many species at risk of extinction, and caused a widespread change in the entire fabric of society in such a large region. Introduction of non-native species to lakes must be prevented with all the political will that governments and the international community can muster. Even large lakes are susceptible to environmental degradation and such degradation can destroy the traditional way of life and sources of food for area residents.

Each of the three basin governments has begun to prepare National Environment Action Plans. Lake Victoria will certainly be a priority element in each nation's plan. However, it is critical that immediate actions be taken to address the overfishing and pollution loading in the basin as well as implementing environmental protection measures in the watershed. An international commission to harmonize water resources and environmental protection institutions and to foster implementation of these actions is desperately needed. While an existing fishery commission among the three nations could be utilized to address fishery concerns, the environmental degradation of Lake Victoria deserves a more comprehensive approach in addressing pollution, ecosystem, and fisheries concerns in a comprehensive rather than a piecemeal approach.

In response to this need, the three governments, through a Tripartite Agreement on the preparation of a Lake Victoria Environmental Management Program, have recently agreed to establish a regional cooperation mechanism for joint preparation and management of a comprehensive program. This effort is being funded by and forms a part of the Global Environmental Facility (GEF), to address the environmental challenges associated with the future of Lake Victoria and its associated natural resources (see Box 3).

Annex 3
Mono Lake in Northeastern California

The System
Mono Lake is a large saline (5000 ppm/L) lake with no outlet located east of the Sierra Nevada Mountains in Northeastern California. The high natural salinity of the lake makes it unique in terms of beneficial use: it provides a very productive habitat for brine shrimp and brine flies which are the food source for many local and migratory birds. The mineral content of the water combined with underground springs have created the attractive "Tufa" landscape. About 150,000 people visit Mono Lake annually to view its beauty, its rich wildlife, and unique landscape, making it an important component of the local recreational economy.

The Problems
In 1940, the City of Los Angeles received legal rights to divert up to 123.4 million m^3 of water from major streams feeding the lake. This water would be used for domestic, industrial and power generation purposes. By 1970, Los Angeles was utilizing the full amount of water it had the right to divert. As water withdrawals continued at this rate, combined with several severe droughts the lake level substantially declined, creating ecological stress. The lake's ecosystem is at great risk. The damage to the lake brought on by the current rate of water withdrawal from the watershed could be irreversible if current water diversion rates are continued.

In 1979, environmentalist groups (for example, Sierra Club) filed a lawsuit against the City of Los Angeles charging it for destroying the lake ecosystem, and requesting that the City reduce its water diversion from Mono Lake. The suit contended that other less environmentally damaging water supply alternatives were available. The water diverted from Mono Lake represented only 17 percent of the water consumed by the City of Los Angeles, making the alternative even more reasonable, even though it would be at a higher cost.

In the lawsuit, the environmentalists provided estimates for the economic value of the ecosystem lost. A study by Loomis (1987), using the Contingent Valuation Method (CVM), estimated the value to society of the Mono Lake ecosystem. He developed three lake-depth and salinity-level scenarios to evaluate effects on beneficial uses including recreational access to the lake, propagation of bird populations, unique habitat function for special bird species, and reduction of dust storms as a result of higher lake levels and their impact on scenic visibility. The three lake levels corresponded to three water diversion scenarios from the watershed. In addition to beneficial use effects, the three water diversion scenarios allowed an estimation of the cost for lost power generation capacity.

Loomis (1987) calculated the total cost of preserving the current Mono Lake water levels to range between $18.7 million to $26.2 million annually. On the other hand, the annual loss of recreational benefits to the state residents by continuing the diversion of water at current rates is estimated to be $1.5 billion. Therefore, from a point of view of social benefit, reallocation of water for maintenance of public trust values at Mono Lake is preferred.

Institutional Resolution

The dispute between the City of Los Angeles and the environmentalist group over Mono Lake's water was not resolved until they reached an agreement at the end of 1993 (Cone 1993). Starting in 1989, the City of Los Angeles stopped diverting water because of the continuous drought. In the same time period, state legislation provided $60 million in state funds for restoring Mono Lake and for the City of Los Angeles to find other sources of water.

Under the agreement, the City of Los Angeles will permanently stop diverting one-third of the water it has historically taken from the streams feeding Mono Lake. Instead, the city will build a $50 million project to distribute reclaimed waste water beginning in 1998. Under the agreement, the city will gradually give up claim for 35,000 acre-feet of Mono Lake water per year. The fate of another 60,000 acre-feet per year of Mono lake water rights the city holds will be decided in 1994 by the State Water Control Board.

Annex 4
Eutrophication Problems in Lake Biwa, Japan

The System
Lake Biwa is the only large inland water body in Japan. The lake has a surface area of 674 km^2 and a maximum depth of 104 m (Mori 1988). It is located near the cities of Nara, Kyoto, and Osaka, and its water drains southwesterly to Osaka Bay via the Yodo River (See Maps Section).

Lake Biwa is the most well known and extensively studied lake in Japan. The lake can be divided into two distinct and different regions: the relatively large and deep northern portion (maximum depth 104 m) and the smaller and shallow southern lake (maximum depth 8 m).

The lake has been a major source of water supply for the city of Kyoto since the late 1800s. Very little industrial activity occurred around the lake until the 1920s, when a major production center for Japan's rayon industry was formed in Otsu on the southern part of the lake. In 1928, wastewater discharges from rayon production seriously affected the catch of a small clam species endemic to the lake (*Corbicula sandai*). The clam was one of the main fishery products from the lake at that time. Protests about the pollution situation resulted in compensation to the fisherman and the eventual construction of improved wastewater treatment by the rayon industry.

Changes in the environmental uses of Lake Biwa and its relative importance to the people living in the lake's basin are demonstrated in Table A4.

The Problems
It was not until the post-World War II economic growth period that pollution problems would resurface in the lake. Rapid eutrophication occurred in the 1960s and early 1970s. During this period, contamination from persistent chlorinated organic compounds (PCB's) also occurred. Accelerated eutrophication has been extensively studied. Problems are most pronounced in the southern lake, which doesn't stratify because of its shallowness. There is aerobic release of phosphorus from the sediments in the summer because of increased photosynthesis and subsequent increases in pH to about pH 9 (Kira 1988b, Kira and Nakamura 1993).

Lessons Learned
The eutrophication in Lake Biwa represents a "classic" lake problem that has occurred in many waterbodies worldwide as a result of population and economic growth in the basin. Rapid urban and industrial growth typically occurred during the mid-twentieth century with little provision for water pollution abatement.

Legislation in the 1970s made an effort to control pollution loading to Lake Biwa. In 1980, a comprehensive ordinance was enacted called the "Lake Biwa Eutrophication Prevention Ordinance." The bill included the following provisions:

- Control of nitrogen and phosphorus in industrial discharges
- Ban on synthetic detergents containing phosphorus
- Implementation of management measures for water and fertilizer use in agriculture
- Proper treatment of livestock wastewater and manure management.

The ordinance was passed largely as a result of public concern over an outbreak of "red tide" in the lake beginning in 1977. Red tide is caused by the abnormal reproduction of a flagellate alga (*Uroglena americana*), that produces an unsightly reddish-brown color in the water and an offensive odor.

The major improvements in nutrient loading to the lake were initially through control of point source loads. Besides legislation, other control measures, including the construction of extensive sewerage networks within the Lake Biwa Comprehensive Development Project (1972-96) have played vital roles in preventing further eutrophication. Banning phosphorus-containing detergents and installing advanced wastewater treatment technology resulted in about 30% removal of N and P in domestic wastewater by the early 1980s. The nonpoint source load reductions were not as rapid, however. There has been a focus on agriculture, which is appropriate, but this is a much more difficult problem to address. Long-standing practices are difficult to modify and some management measures will not work unless properly implemented.

The only long-lasting solution to controlling eutrophication in lakes and reservoirs is to alter human activities to reduce (or preferably eliminate) nutrient loading. This process often takes many years and is sometimes only possible after a major pollution episode that generates public outcry.

Table A4: Evolved Beneficial Uses in Lake Biwa, Japan[1]

	Pre-Historic (-6C)	Ancient (6C-12C)	Medieval (13C-19C)	Feudal (16C-19C)	Modern (19C-1960)	Present (1960-)
Aquatic Resource Use	5	5	5	5	4	3[a]
Water Resources	1	1	1	1[b]	4	5
Transportation	3	3	5	4-5[c]	2-3	2
Pollutant Reservoir	1	1	1	1	3	4[d]
Electricity	0	0	0	0	2-3[e]	3
Scenic	1	3	3	5	4	4
Leisure	1	1	1	1	3	5
Religious/Symbolic	1	4	4	4[f]	3	4-5
Flood Control	1	1	1	4[g]	5	5

[1]On a scale of 1-5 where 5 is the most important activity level

Source: Kada, (1991) and comments by T. Kira (1994)

[a]The importance of fishery products in the economy of Lake Biwa's catchment area is steadily decreasing.

[b]Before the introduction of powerful pumps, riparian inhabitants could not use lake water for daily life. Most villages depended on river or ground water except a small number of settlements on immediate shore lines. The role of Lake Biwa as a source of water started with the construction of an artificial canal from the lake to Kyoto city in 1890.

[c]Transportation on the lake became less important towards the end of the Feudal age with the development of ship routs from Japan Sea to Osaka via Seto Inland Sea, and entirely replaced by railway transportation in the 1890s.

[d]This category is difficult to define. Since lakes serve as a pool of pollutants and nutrients, trapping them in the body of aquatic organisms and bottom mud, they play an important role in cleaning outflowing river water. For example, about 65% of inflowing phosphorus is estimated to be retained in Lake Biwa. This function may not change significantly with time, but its implication becomes more important with increasing pollution/autrification.

[e]Electricity generation did not exist before 1890 when the first hydroelectric power plant started operation in Kyoto using the canal water from Lake Biwa.

[f]There is no evidence that religious activities in and around Lake Biwa were more significant in ancient times than in the feudal age. The symbolic meaning of the lake tends to increase in recent years.

[g]Flood disasters became apparent with the exploitation of lakeside lands into paddy fields, and the development of Osaka city in the lower reaches of the Yodo River that drains the lake water to the sea during the feudal age.

Annex 5
Long-term Water Quality Changes: Effects on Fisheries in Lake Manzala, Egypt

The System
Lake Manzala is a brackish lake/lagoon in the Nile Delta region of Egypt. The lake fishery is in an open water area of 89,000 ha. The open water fishery employs about 17,000 full-time commercial fishermen. This open water area can be divided into four sectors (northern, eastern, southern, and western) that are distinct in terms of water quality, nutrient loading, fisheries effort, and productivity (Toews 1988).

The Problem
The southern sector of open water currently accounts for about 78 percent of the fresh water inflow and 84 percent of the total annual nutrient loading to the lake. This area of the lake has become increasingly eutrophic because of increases in agricultural drainage water inputs. Drainwater inputs have also increased the salinity levels in the lake. The lake area has been reduced by conversion of previously inundated areas to agriculture. The physical changes plus the increased volume of nutrient-rich, agricultural drainage are changing the lake fishery. There has been a gradual transformation from a marine-based fishery to one dominated by fresh water species of relatively low commercial value. This change is not necessarily considered a bad situation, but one that must be managed properly so that it does not adversely affect the ecosystem or commercial fishery.

A summary of the changes in lake area, hydrology, and water quality over time is presented in Tables A5.1 and A5.2. These data are presented separately because the available data were summarized for three slightly different periods (Toews 1988).

Table A5.1: Changes in Area and Hydrology of Lake Manzala, Egypt, Selected Periods 1933-80 (million m^3)

	Period		
Parameter	1933-35	1962-68	1974-78
Lake area, ha	147,000	126,000	89,000
Freshwater inputs	1,399	6,004	6,770
Evaporation	2,565	1,890	1,529
Balance (inputs-evap)	-1,166	+4,114	+5,241

Table A5.2: Changes in Salinity and Nutrients of Lake Manzala, Egypt, Selected Periods 1920-80 (mg/l)

	Period		
Parameter	1920-29	1962-66	1979-80
Salinity, mg/l	14,000	8,000	3,000
Nitrogen, mg/l	No data	0.159-0.634	up to 2.172
Phosphate, mg/l	No data	0.021-0.105	0.050-0.294

The open water area of Lake Manzala is now about half of the 171,000 ha it was in the early 1900s. This change has dramatically altered the hydrologic balance of the lake. During the early 1930s, there was a net evaporation, but the balance changed by the 1960s. Evaporation today is less than one-third the total freshwater input (rainfall plus inflow).

The changes in lake area and freshwater inflows dramatically changed the salinity concentrations in the lake's water. Salinity levels decreased from 14,000 mg/l in the 1920s to 3,000 mg/l by 1979. These changes altered the fishery composition in the lake from a domination by marine species (mainly mullet) to one dominated by freshwater fish (mainly tilapia).

Table A5.3 summarizes the species composition of commercial fish catches in the lake for five periods. The figures in the table show a drastic shift in the fish catch over time

Table A5.3: Summary of Changes Over Time in Species Composition of Catch in Lake Manzala, Egypt, Selected Periods 1920-80

	Period				
Fish Species (tons)	1920-29	1957-58	1962-66	1972-76	1979-80
Marine: Mullet	7,069	3,259	2,820	2,219	902
Crustaceans	2,242	895	2,041	435	2,430
Other Marine	418	968	1,343	347	301
Fresh: Tilapia	2,521	8,894	13,433	19,771	34,822
Eels	393	160	152	225	309
Other Fresh	76	730	1,207	1,094	1,996

Lessons Learned

Several changes in Lake Manzala were brought about mainly through conversion of lands to agricultural production. These changes have altered the fishery in the lake, primarily through dominance of the freshwater species, Tilapia, that outcompeted the marine mullet as the lake water salinity decreased.

The situation points to the need for a comprehensive plan that considers the consequences of major land-use changes on water resources. The fishery changes could have been devastating to the commercial fishery (previously dominated by mullet catches) had the Tilapia not also been a commercially-viable species. Future changes in the watershed that alter lake salinity levels and pollution inputs from fertilizer use, pesticides, sewage, etc. may dramatically and irreversibly effect the commercial fishery and the entire lake-watershed ecosystem.

Annex 6
Eutrophication in Lake Titicaca (Peru and Bolivia)

The System

Lake Titicaca is at an elevation of nearly 4,000 m in the high plateau of the Altiplano on borders between Peru and Bolivia (See Maps Section). It is recognized as the highest large (over 8000 km^2) navigable lake in the world. The area is reputed to be the birthplace of the Inca culture and is the home of indigenous peoples who depend on the lake for survival. Lake Titicaca with its unique aquatic biodiversity, resulting from nearly a million years of isolation, is of special evolutionary interest to ecologists.

The high altitude affects the amount of solar radiation, and the limited exchange of water in its inner bays, make it (particularly Puno Bay) very vulnerable to eutrophication. There is predominantly very low transparency in the lake. The fish community, especially in the inner bay is changing. Native fish occur in smaller numbers, while the introduced perjury is increasing.

Macrophytes (aquatic plants) are represented by at least 15 species, most of which are found in Puno Bay. The macrophyte coverage is of great importance to the indigenous people around Lake Titicaca

Recently, attention has been given to the increasing pollution which affects both humans and the ecosystem. Puno Bay has been particularly affected although it covers only 16 km^2, but supports a population of 90,000 people, including the city of Puno.

The Problems

Many of the observable physical manifestations of lake eutrophication are caused by increased nutrient input, (mainly by nitrogen and phosphorous), resulting from human activities in the watershed. Besides receiving the sewage and industrial waste from the city of Puno, Lake Titicaca receives agricultural run-off from the surrounding areas and tailings from mineral processing plants both in Peru and Bolivia.

Besides affecting the immediate use of water, inadequate wastewater treatment alters the availability of other resources. One of the obvious effects of eutrophication on aquatic organisms is an increase in phytoplankton (microscopic algae), changing the water chemistry, notably in dissolved oxygen. Very low dissolved oxygen levels occur periodically, due to algae die-off, causing severe fish mortality. The fish community thus reacts to the changes in water quality, making more tolerant species dominant. The introduced pejerry feeds on a reduced variety of food items and tolerates the pollution.

Higher pH levels in the inner bay also indicate high photosynthetic production there. The low transparency, due to the excessive algae growth in the eutrophied waters, decreases the macrophytes. The macrophyte coverage has been reduced considerably visibly choked by the algae.

Due to the isolation for millions of years, the zoobenthic fauna of Lake Titicaca shows high degree of endemism. Zoobenthic studies suggest that inner Puno Bay eutrophication has proceeded far beyond an initial phase, and is reaching a state of severe environmental stress. Benthic invertebrate population, which is expected to increase under the initial effects of eutrophication, is much lower in the inner bay than in the rest of the bay. Some of the outer bay taxa is extinct in the inner bay.

Lessons Learned

The large populations of introduced pejerry in the inner bay support local food needs and market fisheries. Coming from an environment heavily loaded with fecal coliform bacteria, however, as well as toxic substances, ingestion of this fish may subject consumers to serious health risks. Work days that may be lost due to food poisoning represents great economic losses for families and a considerable negative factor for the economic development of the Altiplano. Although Puno is the largest city, it is not the only city discharging its wastewater directly to the bay and the development of industrial activities with abundant use of chemical products, will add to the risk of affecting the fisheries of the whole Puno Bay. The Uros people, dwelling among the tortora beds (micro water plants) greatly rely for their survival on this Bay.

In purely economic terms it would be difficult for the lake fisheries to compete withindustry or minimg, and it is therefore important to consider the social aspects for a group of people with few economic resources. In addition to the protein source, fishing income is also important in the household diet and economy. The disappearance of this income would cause many of these shore-dwelling fishermen to emigrate to the larger coastal cities, since their small-holdings (*munifundios*) would be insufficient to sustain the families. It is important to note that although fishing does not generate much income it does channel some income to families who most need it.

The recent decrease in tortora beds may be attributed partly to harder extraction, partly to channel dredging, but the increased eutrophication with excessive algae growth must not be neglected as a cause for the change. Other limiting factors might be chemical contamination. The reduction in macrophyte coverage needs further study.

As environments where many fish and wildfowl species can reproduce, macrophyte beds contribute to the productivity of Lake Titicaca, and thus to the ecological and aesthetic interests of the Titicaca National Reserve, and to its tourism value. On the whole, the macrophytes of

Lake Titicaca and Puno Bay are a resource of exceptional economic and social value for the people of Altiplano and for Peru and Bolivia.

Annex 7
Degradation of the Lake Chad Basin Ecosystem

The System
Lake Chad is one of the larger lakes in Africa (See Maps Section). Its area fluctuates greatly, mostly between 25,000 km^2 and 10,000 km^2. Its drainage area is 25 million km^2. The lake is characterized by the instability of hydrological conditions and by the abundance of macrophytes "Ecologically Lake Chad is closer to the flood zones of large tropical rivers than true lakes," (Daget 1967 as reported in Carmouze et al. 1983). In fact Lake Chad is the second largest wetland in Africa. A major long-term lowering of the lake level took place in 1973, after which occurrence the lake system was referred to as "Lesser Chad". As a result of the lowering the lake became split into three unequal water masses.

Lake Chad and its perennial tributaries are among of the richest fisheries in the world, but only recently has this potential protein resource been recognized. The species composition of the lake is very complex, with 1,305 species of fish. The pre-1973 "Normal Chad" was a tropical lake rich in phytoplankton. Algal biomass increased as the lake area decreased. The composition in the various basins that were created now have to be considered separately.

The environmental changes marked by a combination of the Sahel drought and the accelerated development of fishing activities and other human exploitation, have caused modifications in the fish populations. Better-adapted species have developed but, thriving in the nutrient-rich water, they also carry contaminants.

The Yaeres floodplains of the Logone river, which feeds the lake, support numerous activities such as recession (using moisture betwen floods in floodplain) agriculture, pasture growth, fish breeding and production, Waza National Park tourism, and above all drought fallback security. Lake Chad is situated on one of the major bird migration routes between Europe and the tropical countries of Central and Eastern Africa and serves an important role in the conservation of species.

The Problems
One of the main uses of the lake water is for irrigation in agriculture, making use of its high quality and availability at low cost. The four countries bordering Lake Chad (See Maps Section) have all individually carried out projects using the waters of the lake. To manage the lake, and to promote development, several programs have been set up on national as well as interstate and international levels. The Lake Chad Basin Commission (LCBC) was established in 1964. The

Convention was signed by Cameroon, Chad, Niger, and Nigeria, and supported by United Nations Development Program (UNDP) and United Nations Environmental Program (UNEP).

The Nigerian government established the Chad Basin Development Authority in 1973 in order to control the major irrigation projects in its territory; and a four-year planning and management project of the water resources has been started by UNDP. However, the water level seems to have declined below the threshold of a reversible equilibrium. As the evaporation diminishes, so does the rain; as a result, the desertification process advances and the lake seems to die.

In the Chad basin areas, rural-rural migration occurs over a relatively short distance, following the receding lake, crossing national borders. The lake floor is fast becoming a melting pot for migrants (Kolawole 1989), including farmers, pastoralists and long-distance traders.

The unregulated migration has put pressure on the lake and its basin. Soil degradation has been caused by drought, but also by human activities such as over-cutting of trees for fuelwood or construction purposes, overgrazing, and trampling by livestock, clearing for irrigation, agriculture, waterworks or for human settlements. Degradation of soils has decreased domestic livestock production, floodplain pastures have been lost because of upstream dam projects, causing migration of displaced farmers and pastoralists into marginal lands of the Chad basin area.

In the last decade, the fisheries have suffered from drought, diversion or blockage of instream flows by dams, increased fishery effort by displaced farmers and nomads, smaller mesh sizes, increased juvenile catch, and a growing black market in fish sales. The fisheries have not been monitored since the beginning of civil strife in Chad in the late 1970s.

Although during wet years, the north basin supports more fish than the south (because of its greater depths), during dry years, the north basin dries up first, due to restricted water supply. The stranded fish can not move to the south basin because of an oxygen-low barrier, forming from rotting plant life along the Grand Barrier. The shallowness of the lake furthermore promotes re-suspension of sediments, blocking the gills of fish and reducing their oxygen. Regeneration of stocks is possible as long as floodplain habitat (Yaeres) remains intact and fishing is controlled.

The diminishing of surface water sources in the Chad Basin has intensified groundwater abstraction for domestic and industrial supplies without adequate natural or artificial recharge to the aquifers, an action that may cause land subsidence. Again, humankind is compounding the desertification.

Lessons Learned
The river flow feeding the lake depends on rainfall, over which there is minimal control, and the efficient use of the existing water is therefore of utmost importance. The instream flows should support fisheries in the lake, adequate flooding to supply "natural" irrigation to floodplain pastures, fish breeding sites, agriculture and forest regeneration, and adequate flows to insure the recharge of groundwater.

There can be no separation of conservation from development activities. Single-purpose projects, such as irrigated agriculture, tend to fail or else produce undesirable results. It is necessary to diversify irrigated agriculture by incorporation of tree regeneration and other agroforestry interventions that improve water and soil conservation.

To make better resource management possible, adjustment of the boundaries of the conventional basin is needed and reform of land legislation is necessary in order to encourage local investment and active participation in the resource use. Therefore, overall administrative and legal framework is needed. The heads of state should consider a legally binding framework for regional water resource concerns as suggested in their 1970s meeting.

Although the LCBC is the most relevant international organization, its lack of power has limited the role of arbiter of water- and land-use conflicts within the basin. Environmental planning, monitoring, and assessment is high priority for the LCBC. Heads of state must give LCBC the right to perform environmental and economic impact assessments on any project that might have international consequences. LCBC could be the central organization for monitoring basin-wide changes in the environment, including soil degradation, loss of plant life, wildlife concerns, and water supply. A review of the structure and function of the LCBC and its mandate is needed. The commission should be given appropriate powers to deal with conflicts between the member states in order to avoid armed confrontation over water and land.

Annex 8:
An Environmental Management Program in Lake Baikal Basin

The System
Lake Baikal is located in Russia, near the border with Mongolia. With 23,000 km^3 of fresh watering a surface of 31,500 km^2, it is the largest and deepest freshwater lake (depth 1,620 meters) in the world. The lake accounts for about 80 percent of the reserve of surface freshwater in the former Soviet Union or 20 percent of the entire world's reserve. It was formed 25 million years ago, and has a rich fauna and flora of approximately 2,600 species, of which 59 percent are exclusive to Lake Baikal.

Lake Baikal Ecosystem Problems
Over the last 30 years, about 40 factories have been directly discharging their wastes untreated into the lake. In 1988, three principal sources of direct water pollution, the city of Ulan-Ude (located upstream from the lake on the Selenga River), the Baikal Pulp and Paper Combine (Russia's largest paper plant), and the Selenginsk Pulp and Cardboard Combine, were responsible for a total of 76 percent of the polluted wastewater discharged into the Baikal basin. The city of Ulan-Ude dumped almost 500 tons of nitrates into the water, approximately 70% of all nitrates entering the basin. Baikal Pulp and Paper, the largest single polluter, emitted 32 million cubic meters of wastewater directly into Baikal in 1987. The water near the plant registered above-normal levels of sulfates, chlorides, and suspended particles, and a 20 km^2 tract of lake bottom has been fouled. In addition there are 100 smaller enterprises and settlements located around the lake that have no wastewater treatment capacity at all. Waste flows into the lake from the 336 rivers, of which the Selenga River, flowing out of Mongolia, provides over one-half of the inflow.

Russia has focused attention on this lake because of strong national and international public demands to preserve it. A number of public organizations devoted to the protection of Lake Baikal have been formed. Four joint Communist Party/government resolutions were passed between 1969 and 1987 to clean up and to protect Lake Baikal. The last resolution mandated a development of a long-range integrated plan for the development and protection of the lake. A series of water protection zones in the regions adjacent to the lake that provided strict limits on, and in some areas prohibition of, development. Finally a commission was appointed to oversee implementation of the resolution.

Progress has been seen in the Selenga River. In 1988, there was a 17% decrease in the total volume of waste emitted into the lake from the Baikal Pulp and Paper plant (which scaled down), and a yeast plant was closed. Biological indicators in the surrounding area showed marked improvements in Lake Baikal water quality. Commercial logging operations immediately around the lake have been banned since 1988, and the number of forest fires has been reduced. Even with these efforts, the level of contaminants in the lake is still increasing.

The volume of waste being properly treated increased in the late 1980s, but total volume of untreated waste also increased. Also, in other areas of the lake the levels of suspended particles, petroleum-based substances, and nitrogen compounds discharged to the lake increased.

Reforms to address the lake's condition are falling behind schedule as wastes from untreated effluent continue to be discharged into the lake from some hundred smaller enterprises.

Lessons Learned

Pollution loading has lead to a significant deterioration in Lake Baikal's quality, there is a lingering threat to the lake's ecological viability. Urgent measures are required to define the pollution loading to the lake, to determine the existing health of the system, and to understand the response of Lake Baikal to pollution loading. Lake Baikal is of immense ecological and economic value and deserves attention

An institutional process has been initiated in the Lake Biakal basin, including the passage of a government resolution to prepare a long range plan for development and protection of the lake, and the establishment of a commission to oversee its implementation. The process needs to now move forward with the development of a basin wide plan that includes the participation of all the stakeholders in the basin including government agencies, industry, non-governmental organizations, and the public. Based on recent developments in the region, the public and NGOs appear to understand the seriousness of the situation and the need for action even in the face of difficult economic conditions. The plan needs to identify the critical issues, and to define the order and timing for corrective actions to be implemented.

Progress has been made in redeveloping some measures to regulate industries discharging pollutants to the lake. Toxic substances continue to be a problem with this lake, however. Toxics are concentrating in lake bottom sediments and bioaccumulating in the lakes' biological systems, making them available in unhealthy concentrations to people who eat lake fish. In the case of Lake Baikal, the substances in form of dioxins and furans (which are toxic organic halogenated hydrocarbons), originating from the pulp and paper industries, have been proven to have adverse affects on fish and wildlife, and have been linked to poor human health. The severe reduction or total elimination of these substances needs to be seriously considered as a goal in the planning efforts. Finally, a strong research and investigative program is recommended for Lake Baikal to help define the health of the system more comprehensively, and to determine the best technologies and management practices for improving the lake.

Annex 9
The Aral Sea

The System
The Aral Sea is a brackish water lake located in the arid south-central region of the former Soviet Union. The Aral Sea is fed by the Amu and the Syr Darya rivers originating respectively in Kyrgyz Republic and Tajikistan, and crossing Turkmenistan, Kazakhstan and Uzbekistan, before discharging into the lake. In the 1950s the lake had a water volume exceeding 1,090 km^3, and a surface area of more than 67,900 km^2. The water level in the Aral Sea ranged seasonally between 50-53 meters above sea level.

The Aral Sea had an extremely complex ecological system. It had a dominant moderating effect on the local climate. The mass evaporation from the lake created a screen that kept the micro climate behind it moderate and stable. It protected Central Asia from the cold north winds. Upon meeting the immense column of evaporation, the cold air was lifted to great hights, traveled to far distances in the south and came down to replenish snow deposits and glaciers in the mountains of the Kyrgiz Republic and Tajikistan. Melting snows in these locations were the source for the two rivers flowing back to the lake.

The Aral Sea had an important economic role as a north-south shipping route and as a source of annual fishing catchment of 45,000-50,000 tons of fish. The reed growth along the shores provided the raw material for cellulose and carton production. Sustained pastures and more than 250,000 hectare of tugay forests in the Amu delta, where migrant birds nested and rare animals lived, were a natural barrier against soil erosion.

The Problems
Development of irrigated agriculture during the 1950s in the watershed did not reduce the rivers' run off into the lake, because the areas developed were primarily in valleys and river deltas, areas with abundant water. Sufficient drainage provided appropriate conditions for irrigated agriculture with water consumption of the grown crops being not more than the evapotranspiration of the preceding plants which grew in the area. Consequently, the water balance in the Aral Sea basin was not affected.

In the 1960s, the Soviet government initiated regional irrigation development projects aimed at improving economic conditions in the region. A system of canals and pumps was constructed to withdraw water from the Amu Darya and Syr Darya before their discharge into the Aral Sea, and to convey the water to remote desert areas of Kazakhastan and Uzbekistan. The water was used mainly for irrigation of cotton and rice using very inefficient irrigation technologies. In retrospect, it is astonishing to learn how prominent people viewed the Aral Sea Basin and its watershed as expendable resources for short term economic benefits (Box 8).

As can be seen from Box 8, the central planning approach of the former Soviet Union attempted to maximize physical output from the region at all cost, ignoring environmental damage. As a consequence of increasing water withdrawals and intensified unbalanced agricultural irrigation in the basin, not only the Aral Sea and the diversified ecological system, but also the regional economy have all been dying. Since 1960 the water level has been reduced to 39 meters above sea level (from about 50 m). The volume of the lake has been reduced to 340 m^3 (from 1090 m^3), and the lake area has been reduced to 37,000 km^2 (from 67,900 m^2). The runoff from the rivers into the lake has been as little as 5 km^3 and in some years it has been 0. The salinity of the lake's water in terms of mineral content has been increased from 10 grams/liter in 1960 to 28 grams/liter in 1990.

Box 8. Soviet Official Perspectives on the Aral Sea Region

A. Babayev, former president of the Turkmen Academy of Sciences said: " I belong to those scientists who consider that drying up of the Aral is far more advantageous than preserving it. First in its zone, good fertile land will be obtained. According to preliminary computations, this will provide for 1.5 million tons of cotton a year. Cultivation of this crop alone will pay for the existing Aral Sea with its fisheries, shipping and other industries. Secondly, many scientists are convinced, and I among them, that the disappearance of the sea will not affect the region's landscape" (cited in Precoda 1991, p 111).

Responding to the question why water has a zero cost for agriculture, the Minister of Land Reclamation and Water answered: "Depending on the quality of soil, in irrigating one or another area the need may be for 5,000 or 15,000 m^3 of water per ha so that one activity would for the same task pay three times as much as another. This is not just and simpler to provide water without charge to both."

Source: Precoda (1991) p. 113.

The marked reduction in the lake's area weakened its ability to intercept the Arctic winds. This has caused significant climatic changes in the region including lower temperatures, lesser amounts of snow in the mountains, reduced snowmelt, and less river flow affecting irrigation and recharge to the lake. The significantly lower temperatures in the region and the shortage of water have affected cotton production. Dust storms have caused wind erosion, and deposition of particles containing sodium chloride have caused damages to crops, power lines, concrete structures, and land fertility. Salty dust also causes health problems of the population in the region. Groundwater levels of aquifers in the basin have been lowered by 3 to 8 m, and their water quality has deteriorated significantly. Forests areas have declined, and in many places disappeared. Of the 178 species of animals in the delta of the rivers, only 38 remain. Navigation in the lake is impossible, and fish catches from the lake are null. All the fish for the canning industry, once from the Aral Sea, is imported from other regions.

Restoration or Recovery?

The degradation in the Aral Sea basin has reached the stage of a disaster. Several plans for restoration and partial recovery have been established, all associated with long-term efforts, uncertainties, and huge investments. A system approach is necessary, taking into account all the components comprising the Aral Sea's ecological and socio-economic systems. One of the approaches to restore the region has been reported in Levintanus (1992). It consists of several policy measures and infrastructure investments to be implemented over a period of 20 years. These include: (a) transformation of the cotton-growing and processing sector; (b) development of the chemical industry and water resources saving; (c) reduction of areas under rice; (d) elimination of agricultural produce losses; and (e) development of other sectors of the economy. Among special actions needed, the plan calls for modernization of the irrigation system afforestation and development of mountain horticulture, and (3) reuse and desalinization of the regional drainage flows. The water level in the lake is expected to reach 40-41 meters above sea level (compared with the present level of 39 meters). The water savings associated with the plan has been estimated around 46 km^3 and the capital investment 56 billion rubles in 1990 values (Levintanus 1992).

With the collapse of the Soviet Union in December 1991, and the subsequent independence of Central Asian republics, help from Moscow was dashed. In the absence of major outside aid for solution of the inherited environmental disaster, the five republics need to solve the problem themselves. The environmental degradation and the Aral Sea crisis in Central Asia were not generated locally, nor are the consequences limited to the region only. A solution to these problems must include multifaceted and multi-level programs addressing the socioeconomic and environmental issues of Central Asia. The success of any comprehensive program depends on Central Asian states' political will, regional cooperation, and international aid. Some measures have been included in a major international effort led by the World Bank to help the (Box 9).

Box 9. The World Bank/UNDP/UNEP Plan and Other International Aid to the Aral Sea

The World Bank put forward a preliminary plan to deal with the Aral Sea situation following a Bank delegation's visit to the region in September 1992. The plan was approved in a Washington meeting in April 1993 by officials from the five Central Asian states, international organizations and NGOs. UNEP, and UNDP formulated an integrated program in which 19 specific projects (focusing on needs) were identified. An international fund for the Aral Sea created by the Bank (with a provisional commitment of $260 million by mid-1994) was designed to finance those projects which are supposed to begin in 1996. In addition, the program called for the establishment of an Interstate Council for the Aral Sea (ICAS) to formulate and implement supranational policies. By January 1994, the five states have endorsed the ICAS plan.

Moreover, the Central Asian countries are reaching out to other states for assistance. Collaborative projects on environmental and ecological improvement have been carried out between the Central Asian states and Turkey, the Arab world, Japan, and other countries.

Source: World Bank (1994) and Micklin (1992).

Annex 10
Persistent Toxic Substances: Effects on Intergenerational Human Health and the Environment in the Great Lakes Basin of North America

The System
The five Great Lakes of North America lie on the border between Canada and the United States and are among the largest freshwater lakes in the world. It is no mistake that the Great Lakes Basin has become one of the world's great industrial regions. With its vast water, land, and mineral resources, all the ingredients were present for its trillion dollar economy to develop. As would be expected in an area of heavy industry (producing 20 percent of U.S. and 50 percent of Canadian manufactured goods, including 70 percent of the U.S. steel industry), complex pollution problems developed over the years. Indeed, the U.S. side has about 3,800 direct point source discharges and over 6,000 indirect industrial discharges to municipal sewage treatment systems.

Substantial pollution control progress was made in the 1970s as $10 billion was spent in federal funds on sewage pollution cleanup. Progress was especially critical for Lake Erie, where eutrophication and nutrient overenrichment were reversed during the 1970s and 1980s. However, these pollution control measures have not been sufficient to restore the Great Lakes Basin Ecosystem because of long-lasting contamination with persistent toxic substances that continue to be released from complex pollution sources.

The Continuing Problem
Over a dozen species of Great Lakes mammals, birds, reptiles, and fish are still suffering from high levels of contamination from persistent toxic substances like mercury, cadmium, lead, PCBs, DDT, dioxins, furans, hexachlorobenzene, polyaromatic hydrocarbons, chlorinated hydrocarbon pesticides, and other man-made chemicals from pulp/paper mills, petrochemical plants, steel mills, various manufacturing facilities, and agricultural activities. These toxic substances are long-lived and they can bioaccumulate in living organisms. Cleanup of other pollutants clearly show that these chemicals are associated with very serious and complex reproductive and other health problems in fish and wildlife. Humans and their offspring have also been found to be adversely affected by these chemicals.

New scientific information shows persistent toxic substances common to the Great Lakes to be linked to reproductive, metabolical, neurological, and behavioral abnormalities; immune system suppression, and increased levels of breast and other cancers. Through bioaccumulation in food chains, these man-made chemicals act like artificial hormones to disrupt endocrine systems.

Persistent toxic substances mimic sex hormones, interfere with sexual development, feminize male wildlife, and may be responsible for an epidemic of male and female reproductive system disorders and cancers found in humans. Moreover, persistent toxic substances have been shown in epidemiological investigations in Michigan to be associated with impaired growth, cognitive misfunctioning, and behavioral disorders of children born to mothers who consumed several Lake Michigan fish meals each month during pregnancy. The Great Lakes experience shows that we are not only contaminating the environment and ourselves but that we are also damaging our unborn children as man-made chemicals interfere with fetal development within exposed mothers.

While great strides have been made to reduce the discharge of these chemicals from industries, turning off the water pollution source does not fix the environmental problem. There are deposits of these chemicals in bottom sediments (from past discharges) that must be removed. Concentrations of these chemicals are high in contaminated groundwater near industrial sites and in areas of hazardous waste disposal and dumpsites. In addition, some substances continue to be released into the air (such as mercury from burning coal/oil in electric power generating facilities) and then deposited with precipitation into the lake basins. Stopping point source pollution discharges from industries alone is not enough. Multimedia pathways of long-lasting persistent toxic substances, including old deposits, must be cleaned up. This is very costly, but it provides important lessons. Remedial actions to remove or isolate these toxic substances are now underway in some of the Canadian and U.S. "hotspots" of toxic pollution. In essence, production, use, and disposal must be banned for the worst chemicals to prevent contamination. However, governments hesitated taking expensive remedial cleanup actions during the 1980s because of conflicting data and uncertainties. With this new scientific information, a more aggressive stance has been taken by both countries.

Lessons Learned

Persistent toxic substances must be treated differently than other pollutants--especially in lake/reservoir basins where they can build up in sediment and readily bioaccumulate through food chains to dangerous concentrations. They cause complex reproductive and behavioral changes in the ecosystem and when humans consume the fish or wildlife this exposure can create similar human health problems. Unfortunately, older chlorinated pesticides that are still used in many developing countries, newer triazine herbicides such as atrazine, and surfactants (breaking cell surface) used in plastics and even dishwashing detergents have also found to disrupt human hormone systems. In addition, persistent toxics are now being implicated as a factor in the rising incidence of cancers of reproductive systems (breast and prostate cancers) in highly industrialized countries.

The risks of injuries to humans and the ecosystem of lakes and reservoirs in which the chemicals can bioaccumulate is high if use and discharge of persistent toxic substances continues. The risk of declining learning performance and problem behavior in children and of cancers in adults could have catastrophic effects worldwide in terms of human suffering, industrial liability for that suffering, and associated health costs. These risks should be considered too high to accept. Actions to virtually eliminate the release of persistent toxic substances should be a high priority worldwide. In addition, these risks should be considered to be too costly to accept from an economic perspective. Prevention of chemical pollution is much less costly in the long run than expensive remediation once a problem has been identified.

Approaches in Canada and the United States to control inputs of persistent toxic substances have not, in practice, sufficiently stopped their entry into the Great Lakes environment. Dealing with the thousands of chemicals individually, as both countries have chosen to do, has and will continue to be a never-ending quest. In fact, none of these substances of concern have been virtually eliminated from the environment as called for in the Great Lakes Water Quality Agreement, signed by both countries in 1978. A preventive strategy must be adopted to ensure that persistent toxics are not created in the first place. The onus must be placed on proponents of a chemical's production and use to conclusively demonstrate that the chemical is not harmful to human health or the environment. Many countries, including Canada and the United States, now have such proof of harm resting on governments. This is an impossible burden to shoulder. Discharges of these man-made chemicals must be virtually eliminated and accumulations of them in bottom sediments of lakes and reservoirs must be removed, treated, or isolated so they do not continue entering the ecosystem. Likewise, contaminated groundwater and hazardous waste sites must be isolated so that persistent toxics are not released to the environment. New industries should utilize clean technology and international organizations should make these issues a priority for action to protect the integrity of the human species and to reduce long term costs and irreversible damage to environmental quality.

Annex 11
Remedial Action Plans: Mutual Accountability Using The Ecosystem Approach in the Great Lakes Basin

The System

The Great Lakes Basin (Annex 10), which includs the five lakes Superior, Huron, Michigan, Erie, and Ontario, is heavily used for a variety of water-related activities ranging from drinking and industrial use to transportation, hydropower, and extensive recreation. As described in Annex 1, significant contamination and impairment of parts of the basin has occurred as a result of unacceptable pollution loading. As part of the Great Lakes Water Quality Agreement (WQA) of 1978 (under the direction of the International Joint Commission), a number of programs targeted at specific contaminated sites have been initiated. There are, at present, 43 areas of concern (AOC) in the Great Lakes basin which generally include the major municipal and industrial centers on Great Lakes rivers, harbors, and in connecting channels.

Remedial Action Plan Process

A site can become designated an AOC if it does not meet the objectives of the WQA, thereby impairing one or more of the 14 specified beneficial uses or the area's ability to support aquatic life. These uses range from recreational swimming to supply of clean industrial process water, and from supplied agricultural irrigation water to aesthetic enjoyment. The Great Lakes program has established guidelines or triggering points that identify when the beneficial uses are not being met, making the designation process somewhat systematic. Once designated an AOC, the WQA requires that a detailed Remedial Action Plan (RAP) be developed for the site. The RAP is prepared using a planning process that incorporates a multi-jurisdictional, public participatory, ecosystem-based approach which is described more completely in Box 10 It identifies the sources leading to the degradation, contamination or loss of benefits, and it develops the most comprehensive, time scheduled approach to pollution abatement, remediation, and the return of beneficial uses to the site.

The Remedial Action Process forges links to other local planning processes and encourages meaningful public involvement. It promotes the ecosystem approach by forcing all aspects of the biological system to be evaluated, and accelerates the process of remediation by instituting accountability within and between governments. The Commission is responsible for reviewing and evaluating progress, for resolving pollution issues in the various areas and for recommending additional sites to be designation by the governments. Since 1985, 38 stage one RAPs (problem definition) have been submitted for review and evaluations by the Commission.

The Hamilton Harbor Remedial Action Plan

Hamilton Harbor, a heavily industrialized harbor on the shore of Lake Ontario, has numerous environmental problems affecting its ecosystem (IJC 1991). Liver and skin cancer (neoplasms and epidermal papillomas, respectively) have been reported on several species of fish. Carcinogens are present in harbor sediments. Reproductive anomalies have been observed in snapping turtles. The composition of the benthic (botom dwelling) community is representative of a highly eutrophic and toxic environment. Contaminated sediment must be placed in a confined disposal facility once it is removed from the harbor. Ammonia, phosphorus and nitrogen concentrations exceed the requirements for balanced algae growth in the harbor. Dissolved oxygen levels near the harbor bottom average 0.5-1.0 mg/l in summer. Aesthetics are impacted and costs are incurred by industry who use the harbor water. Accordingly, Hamilton Harbor has been designated an area of concern and the stage one (problem definition) and stage two (remediation) plans have been submitted for review and evaluation.

Box 10. The Ecosystem-Based Approach

A systems approach to analysis recognizes that everything in the system is connected with everything else. In the case of an ecosystem, the links of concern are those that connect human well-being with the quality of the environment. From the environmental perspective, economic production can be described as a transformation process: resources are extracted and harvested from the natural environment and turned into useful commodities for human consumption. In a systems approach, where the law of conservation must apply, a "useful product" must also include the generation of waste residuals and the depletion of natural resources in order to account for a mass material-energy balance. Consumption must similarly be viewed as a physical process where all products are eventually "used up.".

The approach links physical production and consumption processes with the ecosystem. In this framework waste residuals are manifest in three fundamental states: solids, liquids, and gases. The spatial interactions of physical stocks and flows are accounted for, such as air pollutants transported in the high atmosphere and deposited far from their original source. Production processes are considered important factors in restructuring the environment. These processes include, for example urban construction, building roads, and dams. Demographic aspects are considered, not only in terms of population growth and migration, but also in terms of implications of aging populations and changing lifestyles and values. All these factors have various physical effects on the state of the environment. Finally, the dynamic interaction of natural and human activities (i.e. climate, hydrology and geophysical factors) are included in the evaluation.

The ecosystem is viewed in the context of a natural productivity system, that is impaired by human activities. These impairments can be characterized as stresses of the following types: (1) pollution and contaminant loading, (2) harvesting of biological resources, and (3) permanent environmental restructuring. The third impairment reflects the modification of ecosystems to human cultural stresses, that is human settlements, industrialization, and large-scale development projects.

Source: Great Lakes Science Advisory Board (1991).

Hamilton Harbor supports and serves a large iron and steel industry, but other uses of the harbor include recreational boating. It receives many industrial and municipal waste discharges. Through the RAP process, and in consultation with the stakeholders, a wide-ranging, ecosystem-based investigation led to the identification of existing and potential water uses of the harbor. The stakeholders includes environmental groups, industry, municipal, provincial and federal agencies, user groups, and citizens. The stakeholders have also provided guidelines to help define limits on human health effects, sedimentation, public access to the shore, land use planning, discharge from the harbor to Lake Ontario, and the conditions under which waste can be discharged into the harbor.

As a means of ensuring public involvement and input to the process, two additional groups were formed; a management group called the Bay Area Implementation Team (BAIT) and a public consultative organization, the Bay Area Restoration Council (BARC). The BAIT facilitates the implementation and evaluation of remedial actions and consults with the BARC regularly. It also consults other stakeholders on issues related to the implementation of the RAP. The oversight of the RAP process falls under the auspices of the Canada-Ontario Review Board. The duties of the BARC are to ensure protection and appreciation of Hamilton Harbor; coordinate activities with other organizations; solicit and obtain support to implement the objectives of the stakeholder group; assess public perceptions of ecosystem status; and provide advice to governments on restoration of the harbor. The most interesting outcome of this arrangement of the stakeholders is that the BARC, a voice of the public, is incorporated and is managed by a board of directors.

While governments alone do not have sufficient financial resources to restore all of the areas of concern in the Great Lakes Basin, much can be accomplished by pooling resources, increasing coordination and contributing in-kind services to aid in the development and execution of remedial actions. Although the Hamilton Harbor ecosystem is one of the most severely degraded areas in the Great Lakes, there are signs that the degradation is being reversed. Since 1970, reduced pollution loads from specific industrial and municipal waste streams have resulted in gradual improvements in water and sediment quality (Ontario Ministry of Environment 1985). The preparation of a RAP also provided an opportunity for strategically to address the use impairments of the harbor, one of which is fisheries rehabilitation in this ecosystem. Biologically, the abundance and distribution of benthic organisms has improved (Portt et al. 1987) and more than 40 species of fish now inhabit the harbor (Holmes 1988).

Lessons Learned

The Great Lakes Remedial Action Program is a dynamic, inclusive, collaborative process that helps accelerate comprehensive, ecosystem-based assessments of environmental problems in designated areas of concern. The Hamilton Harbor example shows the process time frame can be relatively short in identifying the most urgent needs, and in defining appropriate remedial actions required. It shows the value of stakeholder involvement in program design, for obtaining consensus on issues, establishing the best avenue for progress, and gaining public involvement and support for executing solutions. The RAP process also promotes community participation and decentralization of individual projects. This process is generic, and may be molded to fit other multijurisdictional or multi-stakeholder situations throughout the world. However, issues of sectoral equity and political structure, if not efficiently incorporated into the process, may interfere with achieving desired outcomes.

Annex 12
Eutrophication and Acidification in Scandinavian Lakes

The System
The small and large lakes of Norway, Sweden and Denmark (See Maps Section) are essential natural resources that support economic activities, and provide drinking water for towns and cities. The lakes also play a considerable cultural role.

The Problems
Agriculture, industry, and human sewage are the major causes of eutrophication in Scandinavian lakes. Overenrichment in the waters has led to high levels of algae in both lakes and coastal waters, killing fish, and limiting the recreational. Restoration of the lakes has been successful in some cases, but problems still exist.

Acidification, especially of small lakes, has been manifested through the disappearance of sensitive fish from the lakes. Less than half the sulphur falling as acid rain in Scandinavia is generated locally, while the rest comes from upwind areas of Europe, in particular the United Kingdom, Germany, France, Netherlands, and Central Europe. Much of the toxic depositions in the lakes also come from the same areas in the form of atmospheric deposition..

Lessons learned
The large number of lakes and their important role in the region have created a systematic response by local and federal governments. Several of these experiences are reported here.

Eutrophication in Denmark
In some counties of Denmark, very serious eutrophication is being observed. Freshwater lakes, and saltwater lakes (in coastal zones and coastal waters) are being affected. Intensive use of fertilizer and too much livestock waste have contributed to the contamination of groundwater with nitrates and chemicals. The problems are all linked, and the solutions therefore also have to be linked.

In 1987, the Danish government passed the Action Plan on the Aquatic Environments, meant to reduce nitrogen and phosphorous discharge in the aquatic environments by 50 percent and 80 percent respectively by 1994. With the extensive program, some counties like Funen reduced phosphorous discharge by 80 percent by 1993, but overall nitrogen was reduced only by 10 percent. However, nitrogen load depends on rainfall, and a long time may pass before results are known. It is important to note that in Denmark, unlike Sweden, the reduction in use of nitrogen was entirely voluntary.

Reduction of nutrient can not be achieved through an environment program alone, but must be supported by an agriculture sector program. Over-intensive use of land, too much fertilizer use, and agricultural subsidies linked to agricultural common markets will stand in the way of clean-up of lakes and groundwater in Scandinavia. Major changes in agricultural must be part of the action.

Successful reduction of nutrients in Lake Malaren, Sweden

Lake Malaren, west of Stockholm, is a complicated water system, with many bays of different character. It can be divided into five different basins, each with its own chemical and biological characteristics. The major inflow to the lake is in the west but the inflow in the northwest is also significant. Annual mean temperature in the area is +6 C°. During winter much of the precipitation is in the form of snow, and during the same period parts of the lake are covered by ice.

Population growth around the lake is very high, and it is estimated that one-third of the population of Sweden (8 million in 1994)will live in the drainage area of the lake by year 2000 (Willen 1984). This would further increase the importance of the lake as a freshwater reservoir, and its use for recreation, fishing, transport, and trade. The lake also receives waste water from the population centers and from industries, particularly in the west, and receives nonpoint source pollution from extensive farming.

From 1934 to 1964 there was a five-fold increase in concentration of total phosphorus in the north-eastern part of Lake Malaren. This area has a large city, Uppsala, and the increase was largely due to sewage water (Ahl and Willen 1965). The nitrogen load was particularly high in the western area, where 65% originated from a fertilizer factory. The heavy pollution of the northern areas continued into the 1970s and affected the previously less-polluted middle areas.

The varied nutrient loads in different parts of the lake were also reflected in the plant and animal distribution. The western and northern areas having the largest concentration of phytoplankton, and also considerable amounts of blue-green algae.

Stringent regulations and improved technical capabilities have brought changes to the quality of the water in Lake Malaren. Both improved sewage treatment in the whole drainage area and decreased nitrogen discharge from the fertilizer factory have contributed. In 1965, 65 percent of the population in the drainage area was connected to biological treatment plants, rising to 90 percent in 1970. By 1975, however, combined biological-chemical treatment of sewage water covered 90 percent of the population in the region, although in the rest of Sweden only 50 percent were covered by this combined treatment. The nitrogen discharge from the fertilizer factory in the western basin decreased by 80 percent from 1967 to 1972, with great influence on the water, especially in the western part where the nitrogen concentration decreased by 50 percent (Ahl and Willen 1975).

The sewage reduction measures has led to a decrease of the total phosphorus concentration by almost 50 percent, and also a diminished concentration of chlorophyll (Ch α) from 9.2 to 7.5 mg/m. The transparency of the lake has increased and is similar to that of the 1930s. In the most affected areas, hypereutrophic (areas or objects in a lake) indicators, such as algae, especially the blue-green, have been replaced by such that thrive in moderately eutrophic or mesotrophic (lake conditions between small or no supply of nutrients and over nutrification) environments (Willen 1984).

Removal of phosphorous in Lake Norrviken: Diversion of incoming nutrients

Reduction of nutrient inflow has been successful in Lake Norrviken in southern Sweden with a reduction (87 percent) in phosphorous loading in terms of phosphorous concentration (Ahlgren 1973, 1979). In spite of the improvements, the lake remained eutrophic, because the level of phosphorous was still high. Release of sediment phosphorous has decreased however and should eventually lead to decreased loading. Removal of phosphorous from sewage has been installed in over 95 percent of treatment plants in Sweden. However, this type of treatment is very costly.

Not all lakes will respond favorably to diversion of further nutrients. In these cases the internal nutrient cycles have to be disrupted, for example, by dredging. Removal of a nutrient-rich sediment layer by suction dredging can be effective, although very costly. Obstacles to dredging procedures are the high cost, especially if the lake is of great size, the problem of disposal of the dredge spoils, and the accompanying disruption of the bottom substrate which affects bottom dwelling animals.

Successful removal of phosphorous in Lake Trummen through dredging

In Lake Trummen in central sweden dredging has been successfully carried out. This lake is only about 100 ha and very shallow. Diversion of sewage effluent had not resulted in any significant improvement even after ten years. Studies pointed out that a large proportion of the annual phosphorous supply was coming from sediments (Bjork 1972, 1974). Removal, by suction, of the top one-half meter of particularly phosphorous and nitrogen-rich layer of sediment was carried out to reduce the internal supply of phosphorous. The slurry was treated with alum and the supernatant water with only 30 μg/l phosphorus was pumped back into the lake.

Concentrations of phosphrous and algae declined dramatically, demonstrating that the surface sediments had supplied a very large amount of phosphorous to the overlying water. The lake became slightly deeper after dredging.

Acidification of Swedish lakes

In Sweden, at least 20 000, mainly smaller, lakes are estimated to be affected by acidification or in imminent danger. Fish mortality as a result of acid water was a well-known phenomena as early as the 1900s (Munz 1984).

A previous view that natural soil processes alone could explain extensive acidification has been rejected. It is now accepted knowledge that atmospheric deposition is the main cause of acidification. The chemical composition of the water in acidified lakes is the result of natural and anthropogenic compounds as well as biogeochemical processes in the catchment, and processes in the lake.

The sources of acid are atmospheric deposition (in the form of acid rain), catchment soil, and lake sediments. Hydrological events, particularly snowmelt, have a strong impact on the transportation and concentration levels of acids. Less than half the atmospheric sulphur deposited over Sweden is produced within the country, while the rest of both SOx and NOx emissions (which are derivatives of burned fuel) that contribute to acidification, originates from Western Europe (Johansson and Karlgren, 1974). The main contribution to this atmospheric deposition is the combustion of fossil fuels.

Many properties of lake ecosystems change during acidification. Increases in metal concentrations of aluminum, mercury, lead and cadmium have been found (Dickson 1978). The nutrients availability are affected by internal changes and by changes in the surrounding catchments (Bjork 1974).

Acidification is manifested through the decrease, and (in advanced stages), the loss of the most sensitive fish species. Severe damage, in the form of stunted populations as well as total failure of yearly reproduction, has been observed in numerous lakes (Hultberg 1985) before actual extinction has occured. The most sensitive and most valuable salmon fishes disappear first, but other species like pike and perch successively die off. The most sensitive phases, during the life of the fish, are the roe and fry stages, especially with species spawning in rivers.

Liming of acidified lakes, changes several of the biotic properties in the lakes. There is a rise in pH, alkalinity, and conductivity, and a decrease in transparency and metal concentration. Total phosphorous concentration becomes more or less stable, and after initial adjustment there is an increase in species diversity, biomass, and primary production (Hornstrom et al. 1993). Liming experiments performed in Sweden show that the living conditions for lake biota will be significantly improved although the composition may change. Lime de-toxifies the water and makes possible both the return of sensitive species and eventually a significant increase in species diversity.

Liming was used in Sweden on a small scale, as early as in the 1950s (Fleischer et al. 1993). By 1992, about 6,000 lakes had been limed. The most exposed lakes and those with populations of crayfish and of various salmonid species and of crayfish are now limed, and results show that the pH is often stable for several years.

The annual cost of liming amounts to $25-27 million. Liming operations have been chemically successful and, in many cases considerable biological improvements have been attained (Henriksson and Brodin 1993). Liming could, however, have unwanted effects in those catchments where nitrification is likely to be stimulated by lime addition.

Lake improvements are possible. Ecosystems are resilient and can bounce back to a healthier state if human activities, such as acidification are minimized or eliminated. In a broad sense costs occurred by remediation measures should be recovered in short time through increased recreation and commerce. Biological restoration of lakes by liming, for example, may be adequate in the local, short-term perspective (Olem 1991). In the long term, the elimination of the primary causes is a necessity and a multi-jurisdictional task.

Annex 13
Economic Application Case Studies

A study by Ashworth and Papps (1991) examined the impact of an European Economic Community (EEC) pollution control directive on the profits of firms in the chloride-alkali industry. Although that study does not directly deal with lake pollution, it can very easily be applied, since it is concerned with the problem of a common pool resource. During the seventies, chloride-alkali plants using mercury cells were found in many European countries, such as France, Germany, and Italy, that share joint waterways. In the process of producing chloride, mercury is discharged as waste brine. The EEC mercury directive provides for two possible methods of controlling mercury discharges to water ways. The first method sets a discharge limit on the basis of the plant's capacity, and the second method defines acceptable standards for water quality to be met by all plants. Various definitions of equal treatment are developed and it is argued that knowledge of the relationships between discharges and various methods of abatement is crucial to an investigation of the extent to which the directive meets the equity criteria. Estimation of total, average, and marginal cost implies that these criteria are not met by this directive. The cost (total, average, and marginal) of the mercury directive is not equally allocated using only efficiency considerations. The authors argue that since fairness is not achieved, the value of a "limits approach" policy of the EEC is questionable.

Hipel et al. (1976) applied metagame analysis to resolve water resource problems. The metagame theory is an operations research technique that reconstructs classical game theory on a non-quantitative basis in order to analyze conflicts and to help policymakers to reach an acceptable and equitable economic solution. A major problem in water resources management is how to handle qualitative considerations such as the value of clean water for swimming and boating. Other approaches for quantification of environmental intangibles have been just lately developed, and some still do not provide meaningful and workable results. The metagame analysis can handle both quantitative and qualitative input to a political problem. The parties themselves decide how this input affects their preferences when determining a political solution to the problem. The approach is demonstrated in this paper using a simple water pollution example. The approach has been applied by McBean and Okada (1988) to an acid rain-related conflict between the United States and Canada.

Kilgour et al. (1988) evaluate the load control system for regulating pollution abatement in water bodies, and compare it with the conventional pollution concentration control approach. A load control regulation system, based on Chemical Oxygen Demand (COD), is extremely important for certain closed water systems such as lakes or aquifers. The load control system, is very complicated and somewhat unfair to some polluters since costs of COD reduction are directly affected by the type of pollutant involved. A water management authority can put load control regulations into effect by determining the total allowable COD load for the lake, and then distributing this allowance among the affected polluters in the form of COD quotas. Game theory is applied to an illustrative problem to demonstrate the potential for cooperation between the polluters in order to cope with the overall COD quota.

Annex 14:
International Lake Environment Committee - Survey of the State of World Lakes

The International Lake Environment Committee (ILEC), located in Shiga, Japan, was formed in 1986 as an international NGO and given legal status by the Japanese Environment Agency and the Ministry of Foreign Affairs in 1987. The principal mission of the ILEC is to promote environmentally sound management of natural and man-made lakes and their environments consistent with sustainable development polices. The means chosen by ILEC to pursue their mission is by promoting international research and investigations, by facilitating the exchange of findings and knowledge among experts throughout the world, by publishing newsletters and reports, convening conferences, and by preparing training and guidance materials for lake and reservoir management.

In 1988 and 1990, ILEC produced and published interim reports entitled "Survey of the State of the World Lakes," that contains limnological data and geographic, socio-economic, and environmental information on 117 lakes distributed worldwide, ranging in size from over 6,000 - to less than 2 - km^2 in surface area. Additional volumes have been published in 1991 1992, and 1994. The data acquisition was accomplished by distributing a 16 page, detailed questionnaire, developed by a subcommittee of international experts, and compiled by the ILEC secretariat. The questionnaire, completed for each lake in the data base, contains numerous questions and presents ample opportunity for the individual providing the information to characterize the existing environmental health of the lake, trends in quality, and likely causes for any identified changes. Unfortunately, as has been experienced by most who have attempted to characterize the over-all health of the world's lakes, lack of consistency in sampling, data retrieval, storage and analysis limits the ability to undertake this task. The range in quality of information in the ILEC data set suffers from this same problem, but on the whole, it is a good beginning for a very important need. The project is sponsored by the United Nations Environment Programme, the Lake Biwa Research Institute, other agencies and private companies. The goal of ILEC is to acquire additional lake quality information that will develop and enhance the data base, compile the information and continue to publish additional volumes for distribution.

Other products that have been produced by ILEC in cooperation with UNEP that would be of interest to country officials, resource managers, and scientist and engineers, interested in protecting or enhancing the quality of lakes, consist of a series of guidelines on lake a reservoir management including: *Principles of Lake Management* (1988), *Lakeshore Management* (1990), *Socio-economic Aspects of Lake/Reservoir Management* (1991) and *Toxic Substances in Lake Management* (1992).

Those interested in obtaining additional information on the ILEC, its work, and its information can contact the organization at 3-13-25, Kusatsu, Kusatsu-city, Shiga 525, JAPAN. Tel: 81-775-67-2155.

References

Ahl, T. and Willen, T. 1965. "The Lake Malaren Research: an Introduction." *Sven. Naturvetensk*: (English Summary), 301-16. Cited in: F. B. Taub, ed., *Ecosystems of the World, 23: Lakes and Reservoirs*, Chapter 5, The Large Lakes of Sweden: Vanern, Vattern, Malaren, and Hjalmaren. Elsevier: Amsterdam, 1984.

Ahl, T., and E. Willen. 1975. "Lake Malaren." *Eutrofiering*. NORDFORSK *Publ.* (1): 85-98.

Ahlgren, I.. 1973. "Role of Sediments in the Process of Recovery of an Eutrophied Lake." In M.L. Golterman, ed., *Interaction between Sediments and Fresh Water*, UNESCO: Paris 372-377.

Ahlgren, I.. 1979. "Lake Metabolism Studies and Results at the Institute of Limology in Uppsala." In Welch, E.B., ed., *Ecosystems of the World, 23: Lakes and Reservoirs*, Chapter 23, *Lake Restoration Results: The Large Lakes of Sweden: Vanern, Vattern, Malaren, and Hjalmaren*. Elsevier: Amsterdam, 1984.

Ashworth, John and Ivy Papps. 1991. "Equity in European Community Pollution Control." *Journal of Environmental Economics and Management* 20:46-54.

Atkinson, Scott, and Tom Tietenberg. 1991. "Market Failure in Incentive-Based Regulation: The Case of Emissions Trading." *Journal of Environmental Economics and Management* 21(1):17-31.

Bardach, John E. 1988. "Riverine and other Tropical Lakes and Their Fisheries." In N. W. Schmidtke, ed. *Toxic Contamination in Large Lakes*. Vol. IV. Chelsea, Mich.: Lewis Publishers.

Baumol, William J., and Wallace E. Oates. 1992. *The Theory of Environmental Policy*. Cambridge University Press, Cambridge, Mass.

Becker, Nir, and K. William Easter. 1995. "Dynamic Games on the Great Lakes." In Ariel Dinar and Edna Loehman, eds, *Water Quantity/Quality Management and Conflict Resolution*. Westport, Conn.: Praeger.

Beadle. L. C.. 1981. "Limonology of African Inland Waters." In *The Ecology and Utilization of African Inland Waters*. Nairobi: United Nations Environment Programme.

Bjork, S. 1972. "Exosystem studies in connection with the restoration of lakes." Verh. Int. Ver. Limnol., 18: 379-87. In *Lake Restoration Results*, Welch, E.B. 1972.

Bjork S. 1971. European Lake Rehabilitation Activities. Institute of Limnology Report, University of Lund. In *Lake Restoration Results*, Welch, E.B., 1974.

Braden, J. B., and C. D. Kolstad, eds. 1991. *Measuring the Demand for Environmental Quality*. North Holland, New York, NY.

Cann, Bud. 1989. "Take it up with the board! (Water Pollution at Lake Hopatcong, New Jersey)" *EPA Journal* 15(6):24-25.

Carmouze, J.-P, J-R. Durand, and C. Leveque, eds. 1983. *Lake Chad: Ecology and Productivity of a Shallow Tropical Ecosystem*. The Hague: Dr. W. Junk Publishers/Kluwer Academic Publishers Group.

Cone, Marla. 1993. "DWP Agrees to Take Less Mono Lake Water." *Los Angeles Times*, December 14.

Cooke, Dennis G., Eugene B. Welch, Spencer A. Peterson, and Peter R. Newroth. 1993. *Restoration and Management of Lakes and Reservoirs* (second edition). Boca Raton, Florida, Lewis Publishers.

Covich, A. P. 1993. "Water and Ecosystems." In Peter H. Gleick, ed. *Water in Crisis: A Guide to the World's Fresh Water Resources*, Oxford: Oxford University Press, Oxford.

Dickson, W., 1978. "Some Effects of the Acidification of Swedish Lakes." Proceedings of the International Association of Applied Limnology 20:851-6. In Acid Water research in Sweden - Knowledge for tomorrow?: *Ambio*, 22(5). 1993.

Dickson, W. 1980. "Properties of acidified waters." In D. Dralbos and A. Tollan, eds. *Ecological Impact of Acid Precipitation* Oslo: SNFS.

Dinar, A., and D. Zilberman. 1991. "The Economics of Resource-Conservation, Pollution-Reduction Technology Selection: The Case of Irrigation Water." *Resources and Energy* 13: 323-48.

Downing, Paul B. 1984. *Envionmental Economics and Policy*. Boston, Mass.: Little, Brown and Company.

Eilers, J. M., G. J. Lien, and R. G. Berg. 1984. "Aquatic Organisms in Acidic Environments: A Literature Review.", Technical Bulletin no. 150. Madison: Wisconsin Department of Natural Resources.

Endter, Joanna Lynne. 1987. "Cultural Ideologies and the Political Economy of Water in the U.S. West: Northern Ute Indians and Rural Mormons in the Uintha Basin, Utah." Ph.D. Dissertation, University of California, Irvine.

Fleischer, S., G. Andersson, Y. Brodin, J. Dickson, J. Hermann, and I. Munz. 1993. "Acid Water Research in Sweden - Knowledge for Tomorrow?" *Ambio* 22(5):258-63.

Ganrot, P. O. 1986. "Metabolism and Possible Health Effects of Aluminum." *Environmental Health Perspectives* 65: 363-441.

Gasschignard, O., and A. Berly. 1987. "Effects of Large Discharge Fluctuations on the Macroinvertebrate Population Downstream of a Dam." In J.F. Craig and J.B. Kemper eds. *Regulated Streams, Advances in Ecology*. New York: Plenum Press.

Golubev, G. N. 1989. "Lake Management and Sustainable Development." In Salanki, J. and S. Herodek, eds. *Conservation and Management of Lakes, Lake Management and Sustainabe Development*. Budapest: Akademiai Kiado.

Great Lakes Science Advisory Board. 1991. 1991 Report to the International Joint Commission, Washington, D.C. and Ottawa: IJC.

Hecky, R. E. 1993. "The Eutrophication of Lake Victoria." *Verhandlungen International Verein. Limnology* 25:39-48, 1993.

Hecky, R. E., E. J. Fee, H. J. Kling, and J. W .M Rudd. 1981. "Relationship between primary production and fish depletion in Lake Tanganyaka" *Transactions of the American Fisheries Society*, 110, 64-71.

Hecky, R. E., D. J. Ramsey, R. A. Bodaly, and N. E. Strange. 1991. "Increased Methylmercury Contamination in Fish in Newly Formed Freshwater Reservoirs." In T. Suzuki et. al., *Advances in Mercury Toxicology*. New York: Plenum Press.

Hipel, Keith W., R. K. Ragade, and T. E. Unny. 1976. "Political Resolution of Environmental Conflicts." *Water Resources Bulletin,* 12 (4):813-27.

Hitzhusen Fred, Somskaow Bejranonda, Timothy Lehman, and Robert Macgregor 1995. "Economic and Political Analysis of Dredging Ohio's State Park Lakes." In Ariel Dinar and Edna Loehman, eds. *Water Quantity/Quality Management and Conflict Resolution.* Westport, Conn.: Praeger.

Hultberg, H. 1985. "Changes in Fish Populations and Water Chemistry in Lake Gardsjon and Neighboring Lakes during the Last Century." *Ecology Bulletin* 37: 64-72.

Hornstrom, E., Ekstrom, C., U. Miller ,and Dickson, W. 1973. "Effects of the Acidification on Lakes in the Swedish West Coast Region Phytoplankton, Zooplankton, Diatoms, Chemistry." In: Acid Water Research in Sweden - Knowledge for Tomorrow? *Ambio* 22(5):258-263, 1993.

Henriksson, I., and Brodin, Y. W. 1993. "How to Save Acidified Waters - A Synthesis of the Swedish Liming Programme. In Acid Water Research in Sweden - Knowledge for Tomorrow? In *Ambio* Vol 22 No. 5, August.

IJC (International Joint Commission / United States and Canada) 1991. *Review and Evaluation of the Great Lakes RAP Remedial Action Plan Program 1991.* ISBN 1 -895085-25-X, Washington, D. C. and Ottawa, Published by IJC.

Kada, Yukiko. 1991. "Understanding the State of the Lake Environment from a Socio-Cultural Perspective - An Example from Lake Biwa, Japan." In M. Hashimoo, ed. *Guidance on Lake Management* Vol. 2. Shiga, Japan: The International Lake Environment Committee Foundation and thc United Nations Environment Program.

Kilgour, Marc D., Norio Okada, and A. Nishikori 1988. "Load Control Regulation of Water Pollution: An Analysis Using Game Theory." *Journal of Environmental Management* 27:179-94.

Kira, Tatuo. 1988a. "The Historic Role of a Large Lake in Japan - The Case of Lake Biwa." In Schmidtke, N. W., ed. *Toxic Contamination in Large Lakes.* Vol. IV, Chelsea,Michigan: Lewis Publishers.

Kira, Tatuo. 1988b. "Some Aspects of Ecological Watershed Management for the Control of Eutrophication, The Case of Lake Biwa Watershed." *Water Resources Development,* 4(4):259-69.

Kira, Tatuo. 1993. "Major Environmental Problems in World Lakes." *Memorie,* 52:1-7.

Kira, Tatuo, and Masahisa Nakamura. 1993. "Environmental Management of Lake Biwa and its Catchment Area with Special Reference to the Control of Eutrophication." In Sund Horst, Hans-Henning Stabel, Walter Geller, Yu Xiaogan, Yuan Kechang, and She Fening, eds. *Environmental Protection and Lake Ecosystems.* Nanjing: China Science and Technology Press.

Kira. Tatuo. 1994. Personal Communication. Letter of comments dated December 15, 1994 to A. Dinar.

Le Marquand, David. 1989. "Developing River and Lake Basins for Sustained Economic Growth and Social Progress." *Water International* (May):127-38.

Le Marquand, David. 1993. The International Joint Commission and Changing Canada-United States Boundary Relations. *Natural Resources Journal* 33 (Winter): 59-91.

Le Moigne, Guy, Ashok Subramanian, Mei Xie, and Sandra Giltner, eds. 1994. *A Guide to the Formulation of Water Resources Strategy.* Technical Paper no. 263, Washington, D.C.: The World Bank.

Levintanus Arkady. 1992. "Saving the Aral Sea." *Water Resources Development*, 8(1):60-64.

Loehman, E., and A. Dinar. 1994. "Cooperative Solution of Local Externality Problems: A Case of Mechanism Design Applied to Irrigation." *Journal of Environmental Economics and Management* 26:235-56.

Loomis, J. B. 1987. "Financing Public Trust Resources of Mono Lake and Los Angeles." *Water Resources Research* 23:1449-56.

McBean, Edward A., and Norio Okada. 1988. "Use of Metagame Analysis in Acid Rain Conflict Resolution." *Journal of Environmental Management* 27:153-62.

McConnell, K. E. 1990. "Double Counting in Hedonic and Travel Cost Models." *Land Economics* 6(2):121-27.

McDonald, M. E. 1985. "Acid Deposition and Drinking Water." *Environmental Science and Technology* 19: 772-76.

Micklin, Philip P. 1992. "The Aral Sea Crisis: Introduction to the Special Issue." *Post Soviet Geography* 33(5):269-70.

Moore Molly. 1994. "Kashmir Fighting Speeds Pollution of India's Idyllic Dal Lake." *The Washington Post*, November 1, p. A15.

Muniz, I. P. 1984. "The Effects of Acidification on Scandinavian Freshwater Fish Fauna." *Philosophical Transactions of thae Royal Society of London* no. B305, 517-28.

Muniz, I. P., and H. Leivestad. 1980. "Acidification - Effects on Freshwater Fish." in Dralbos and Tollan, eds. *Ecological Impact of Acid Precipitation.* Norway.

Northcote, T. G., S. P. Morales, D. A. Levy, and M. S. Greaven, eds. 1989. *Pollution in Lake Titicaca, Peru: Training, Research and Management.* Vancouver: Westwater Research Center.

Portt, C. B., V. W. Cairns, and C. CK Minns. 1987. "Benthic macroinvertebrates and sediment characteristics of Hamilton Harbour Canada." *Tech. Rep. Fish. Aquat. Sci.*

Precoda, Norman . 1991. "Requiem for the Aral Sea." *Ambio* 20(3-4):109-14.

Randall, Alan. 1981. *Resource Economics*, Columbus, Ohio: Grid Publishing Inc.

Raskin P., E. Hansen, Z. Zhu, and D. Stavisky. 1992. "Simulation of Water Supply and Demand in the Aral Sea Region." *Water International* 17:55-67.

Rogers, P. W., D. R. Klomars and P. L. Freedman. 1988. In N.W. Schmidtke, ed. *Toxic Contamination of Large Lakes*, Vol. III. Chelsea, Michigan: Lewis Publishers.

Russell, C., and J. Shogren, eds. 1993. *Theory, Modeling, and Experience in the Management in Nonpoint-Source Pollution*. Boston, Mass.: Kluwer Academic Publishers.

Schmidtke, Nobert W., *Toxic Contamination in Big Lakes*. Vol IV. Lewis Publishers, Chelsea, Michigan, 1988.

Stark, Robert C., David S. Bullock, and Wesley D. Seitz. 1995. "Political Power Measures for Water Management Interest Groups." In Ariel Dinar and Edna Loehman, eds. *Water Quantity/Quality Management and Conflict Resolution*, Westport, Conn: Preager.

Smith, V. Kerry. 1993. "Nonmarket Valuation of Environmental Resources." *Land Economics*, 69(1):1-26.

Swedish Ministry of Agriculture, 1982.

Tietenberg, Tom. 1988. *Environmental and Natural Resource Economics*. 2nd ed. Glenview, Ill.: Scott, Foresman and Company.

Thornton, Jeffrey A., and Walter Rast. 1994. "Not American Lake Management," *Lakeline* 14(3). 11-14.

Tolba, Mustafa K.. 1984. "Caring for Lakes in the Developing World." *Proceedings of the Shiga Conference on Conservation and Management of World Lake Environment*, August 23-31, Japan.

Toews, D. R. 1988. "Fishery Transformation on Lake Manzala, Egypt During the Period 1920 to 1980." In N. W. Schmidtke, ed. *Toxic Contamination in Large Lakes*, Chelsea, Mich.: Lewis Publishers.

UNDP (United Nations Development Programme). 1994. Lake Victoria Environmental Management Programme, Minutes of the Intergovernmental Meeting held on 27-29 April in Nairobi. Dated May 11, 1994.

USEPA (United States Environmental Protection Agency). 1988. *The Lake and Reservoir Restoration Guidance Manual*, EPA 440/5-88-002, February. Washington, D.C.: Government Printing Office.

USPHS (United States Public Health Service). 1965. Report on *Pollution of Lake Erie and its Tributaries*. U.S. Department of Health, Education and Welfare. Division of Water and Pollution Control. Washington, D. C.: Government Printing Office.

Vickrey, William S. 1992. "Theoretical and Practical Possibilities and Limitations of a Market Mechanism Approach to Air Pollution Control." *Land Economics* 68 (1):1-6.

White, James C., ed. 1985. *Liming Acidic Waters-Environmental and Policy Concerns*. Rochester, New York: Center for Environmental Information.

Willen, Eva. 1984. "The Large Lakes of Sweden: Vanern, Vattern, Malaren and Hjalmaren." In F.B. Taub, ed. *Ecosystems of the World*. Elsivier.

World Bank 1993. *World Development Report 1992*. Washington, D.C.

World Bank. 1994. "Aral Sea Program - Phase 1." Briefing Paper for the Proposed Donors Meeting on June 23-24, Paris, 1994. Europe and Central Asia Region, Aral Sea Program Unit, World Bank, Washington, D.C.

WRI (World Resources Institute) 1994. *World Resources 1992-93*. Washington, D.C.

Maps Section

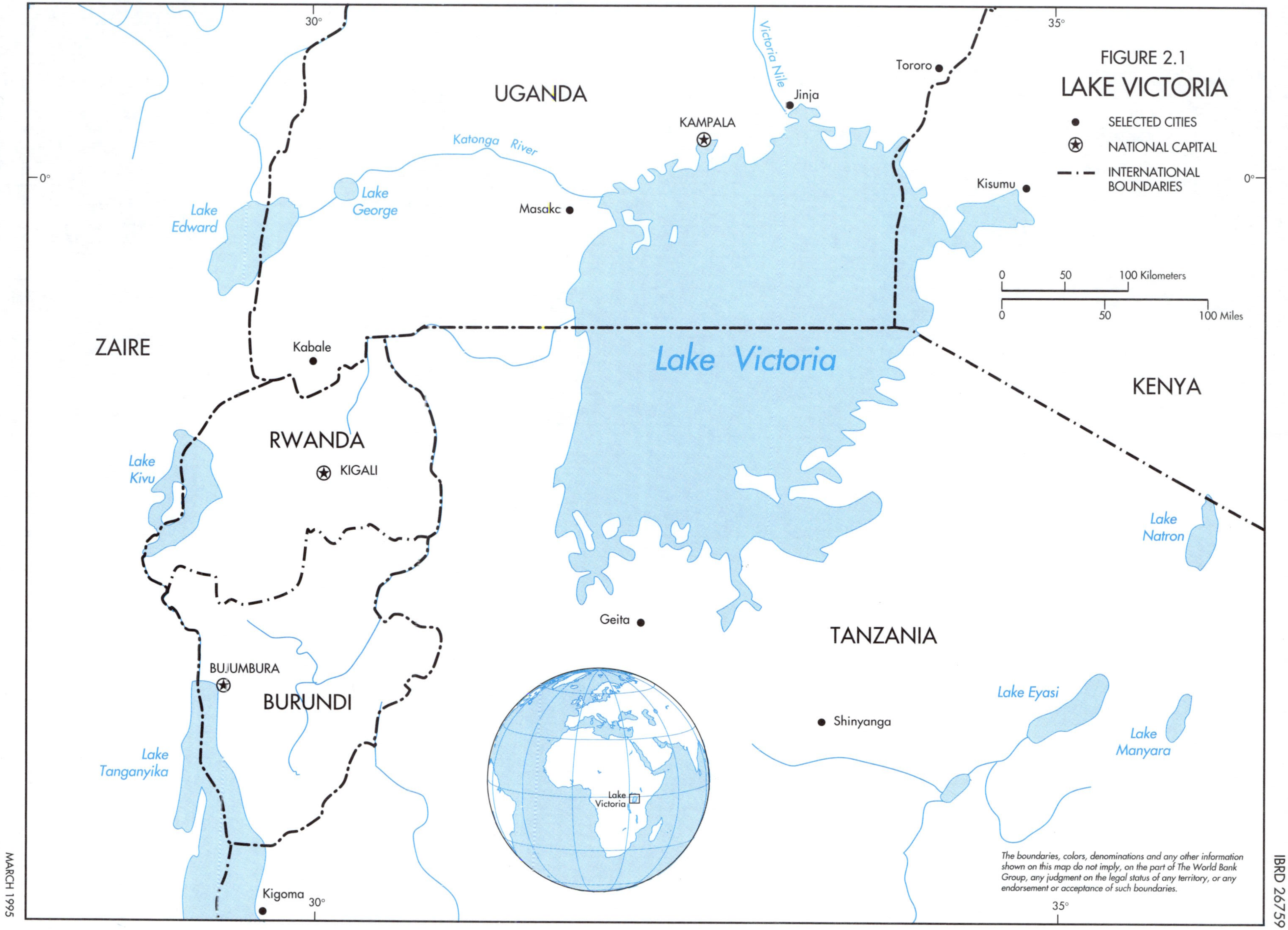
FIGURE 2.1
LAKE VICTORIA
SELECTED CITIES
NATIONAL CAPITAL
INTERNATIONAL BOUNDARIES
0 50 100 Kilometers
0 50 100 Miles
UGANDA
KENYA
TANZANIA
ZAIRE
RWANDA
BURUNDI
Lake Victoria
KAMPALA
Jinja
Tororo
Kisumu
Masakc
Kabale
KIGALI
BUJUMBURA
Kigoma
Geita
Shinyanga
Victoria Nile
Katonga River
Lake George
Lake Edward
Lake Kivu
Lake Tanganyika
Lake Natron
Lake Eyasi
Lake Manyara
Lake Victoria
30°
35°
0°
The boundaries, colors, denominations and any other information shown on this map do not imply, on the part of The World Bank Group, any judgment on the legal status of any territory, or any endorsement or acceptance of such boundaries.
IBRD 26759
MARCH 1995

IBRD 26762

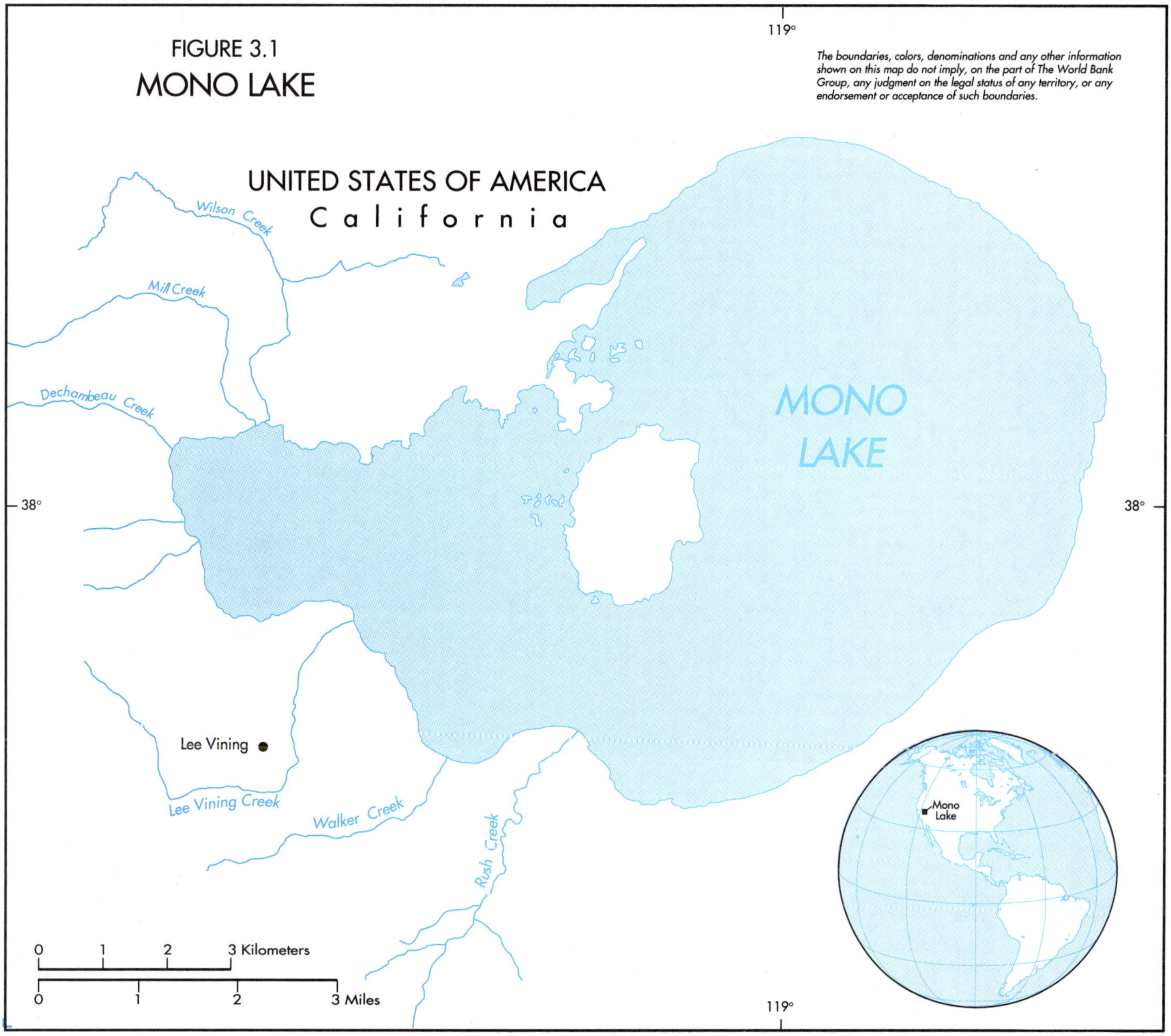

MARCH 1995

IBRD 26763

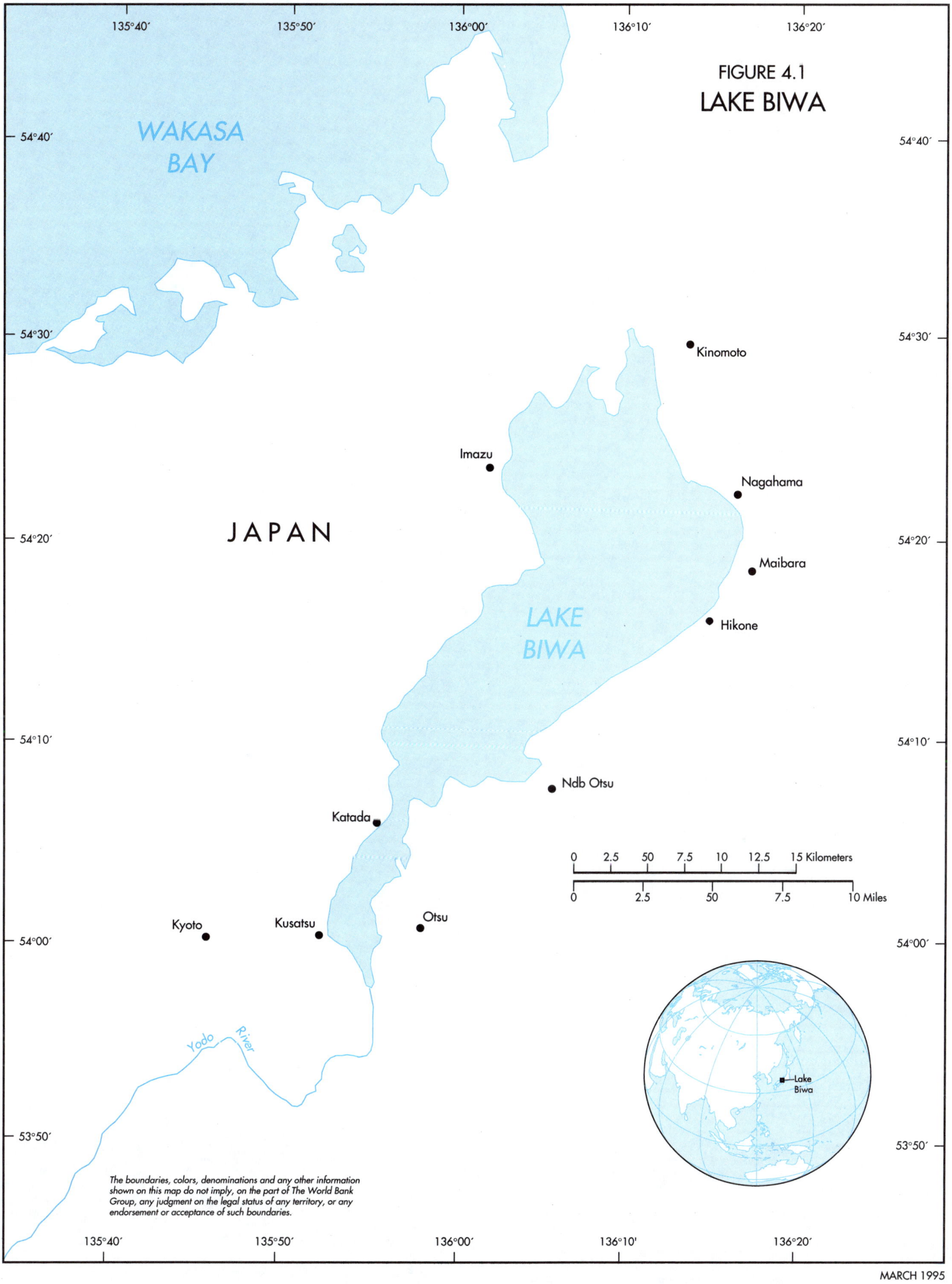

FIGURE 4.1
LAKE BIWA

MARCH 1995

IBRD 26764

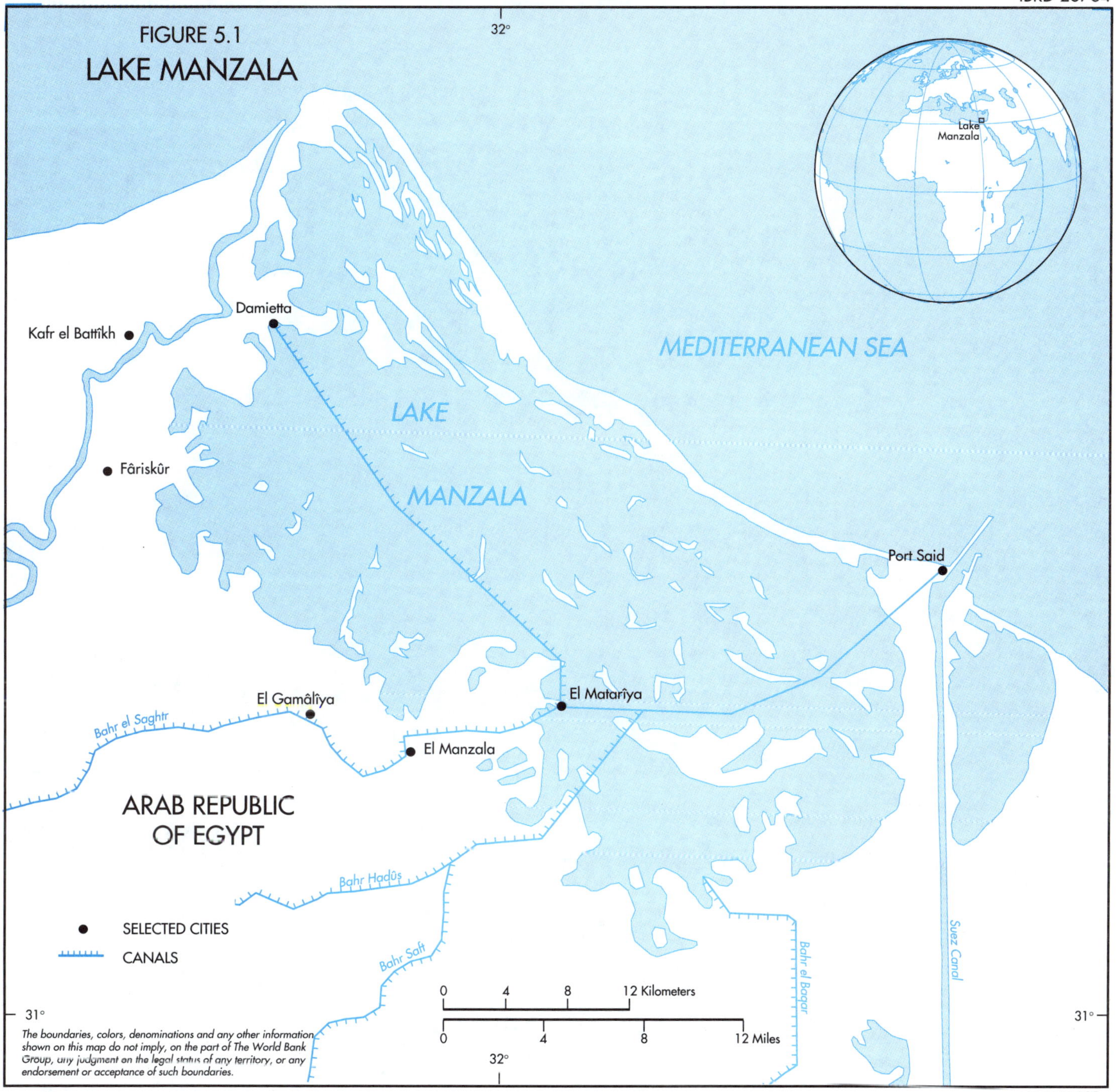

MARCH 1995

IBRD 26765
The boundaries, colors, denominations and any other information shown on this map do not imply, on the part of The World Bank Group, any judgment on the legal status of any territory, or any endorsement or acceptance of such boundaries.
FIGURE 6.1
LAKE TITICACA
SELECTED CITIES
INTERNATIONAL BOUNDARY
70°
69°
15°
16°
Rio Pucaro
Rio Cabanillas
Juliaca
Puno
LAKE TITICACA
BOLIVIA
PERU
Achacachi
Copacabana
Aygachi
Guaqui
Jesús de Machaca
Rio Huenque
Rio Desaguadero
Lake Titicaca
0 10 20 30 Kilometers
0 10 20 30 Miles
MARCH 1995

IBRD 26766

MARCH 1995

IBRD 26760

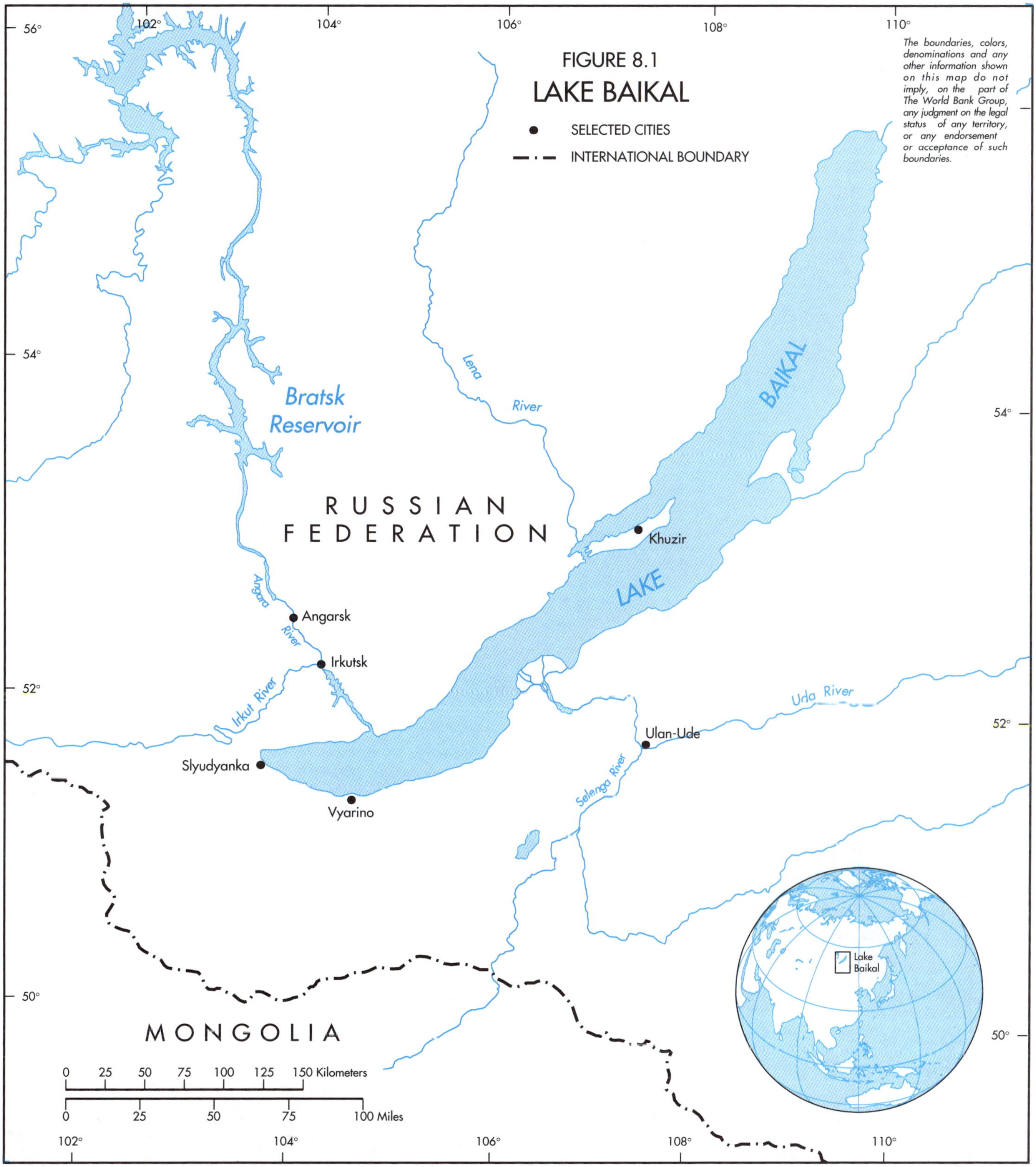

MARCH 1995

IBRD 26761

MARCH 1995

IBRD 26758

FIGURE 10.1
THE GREAT LAKES BASIN

MARCH 1995

IBRD 26767

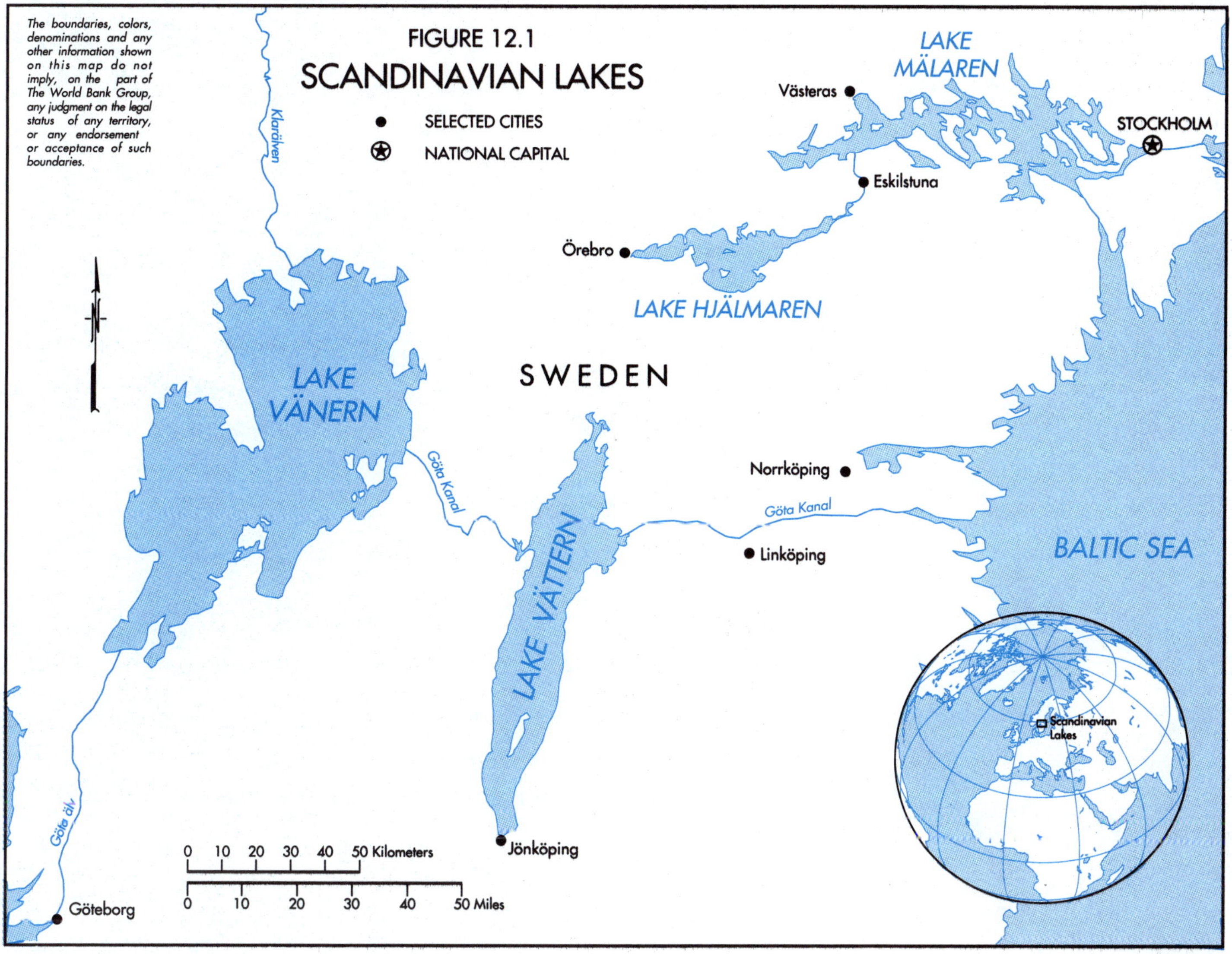

MARCH 1995